FARM WELDING

Andrew Pearce

Farming Press

First published 1992

ISBN 0 85236 230 7

A catalogue record for this book is available from the British Library

Published by Farming Press Books, Wharfedale Road, Ipswich IP1 4LG, United Kingdom

Distributed in North America by Diamond Farm Enterprises, Box 537, Alexandria Bay, NY 13607, USA

Cover design by Andrew Thistlethwaite
Cover photographs by Peter Adams
Typeset by Galleon Photosetting, Ipswich
Printed and bound in Great Britain by Butler & Tanner Ltd, Frome and London

FARM WELDING

CONTENTS

INTRODUCTION

Almost anyone can weld. All that's needed is basic hand/eye coordination, an appreciation of the processes involved and a little guided practice.

But nobody as yet has come up with a book that grabs the reader's hands and says 'Do it like this!' Pictures, though, are worth a lot of words. What follows is visual backup for beginners or a fault-finding guide for those who want to improve. It's the distillation of many Agricultural Training Board welding courses run by the author, plus techniques learnt during some considerable time spent making and mending equipment on a mixed farm.

While not intended to be the last word on welding techniques—plenty of other books have far better claim to that—the advice here is practically based and aimed straight at the on-farm welder. Subjects are broken into sections; each is a mix of words and pictures. Captions are sometimes lengthy and may go beyond what's in the picture, so it's worth reading all that's there.

First, the basics—just what is welding? It's the process of joining materials, usually by raising the joint temperature so components either melt together or fuse under pressure.

The blacksmith traditionally used both heat and pressure when forge welding—heating the joint components, then fusing them by hammer blows—but here we're concerned with letting heat alone do the work. It's supplied either by an electric arc (in MMA or MIG/MAG work) or a gas flame (in gas welding and brazing). Extra metal is usually added to the joint from an electrode or filler rod.

Plastics too can be welded, but they're beyond our brief.

So what's coming? The book splits into sections, with the most common technique (arc welding mild steel) covered first. Equipment and consumable selection, plant set-up and use in various positions are looked at. And, as most people are beset by the same problems, common faults are analysed and help suggested. This approach is followed throughout.

Next comes the farmer's flexible friend—gas welding and cutting. Then it's the turn of the new boy: MIG/MAG welding, equipment for which is rapidly finding its way into workshops. There's a section on soldering, and finally, as backup to developing welding skills, a few basic blacksmithing techniques are shown. It's surprising how often the ability to shape metal with heat and a hammer complements ways of joining it.

Books notwithstanding, the best way to improve is via a good course. Local colleges often run these over the winter; alternatively, the ATB can provide initial or advanced instruction via on-farm or college-based courses.

This book is based on articles which first appeared in *Power Farming* magazine between 1984 and 1990. Welded examples pictured were produced by the author, by the Welding Institute's ex-chief instructor Max Rughoobeer or come from the picture files of the Welding Institute. Blacksmithing techniques were demonstrated by Sussex smith Frank Dean. The help of all is greatly appreciated.

ANDREW PEARCE
April 1992

Section 1

MANUAL METAL ARC WELDING

Manual Metal Arc (MMA or 'stick') welding is still the most common technique on the farm. We kick off with a look inside and outside the welding set, then move on to primary safety (Figs 1.1–1.10).

Identifying a good weld comes next, followed by what's actually happening at the rod tip (Figs 1.11–1.13).

To finish off, operator relaxation and suggestions to make arc striking easier are covered (Figs 1.14–1.17).

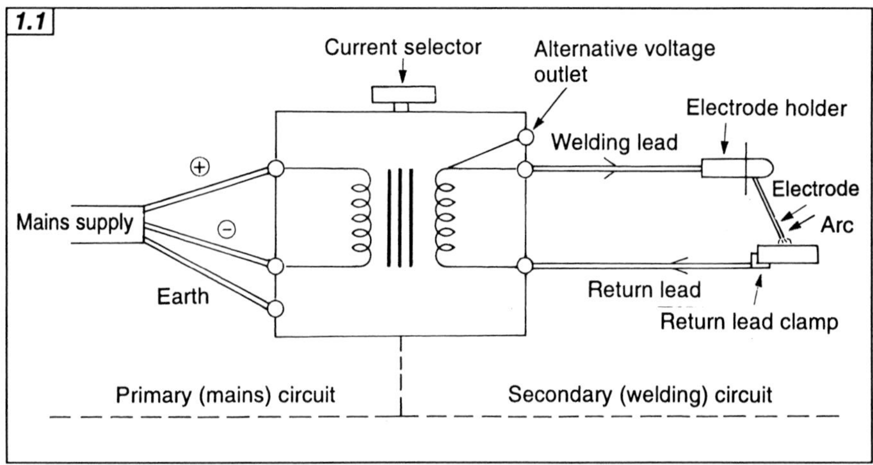

1.1 Where does welding heat come from? The set is a simple transformer. High voltages and low currents go in (left); low voltage(s) and high currents come out (right). Both DC (direct current) and AC (alternating current) output sets are produced. The AC output model, which is cheaper to build and works perfectly well, is easily the most common. Inside the set, single- or three-phase mains power is commonly stepped down to around 50V, though some sets offer higher alternative outputs up to a legal maximum of 100V. Maximum current depends on the set; for general farm use, something delivering up to 200A is fine. Bigger sets will be working well within themselves at lower currents, but are much more bulky, hard to move and usually require a three-phase supply. For non-stop work, go for oil-cooled gear. All sets have output current adjusted by a stepped or continuously variable selector (see Fig 1.3).

Note: Welding current can be pictured as flowing out of the transformer to the electrode holder, down the electrode, across the arc and back via the welding return lead (arrows). It's tempting to call the return lead an 'earth', but that term applies only to the incoming supply's green/green-yellow earthing wire. Don't mix them up; the supply's earth keeps the user safe in case of short-circuit—see Fig 1.6.

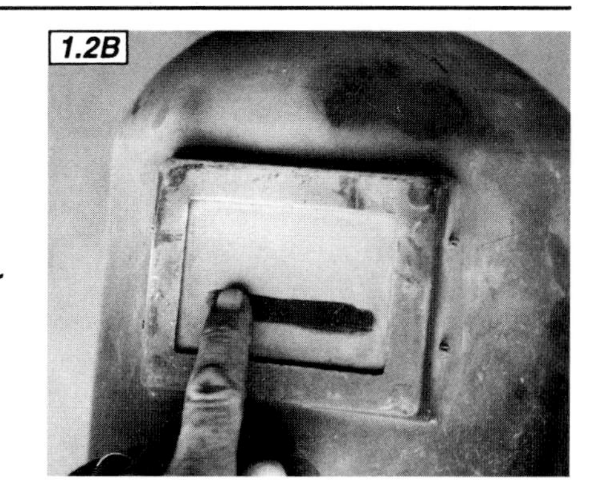

1.2 With welding filters, BS 679 is your guarantee of safety (A). Use only something with this mark or with DIN 4647 part 1 (1977). Why? The arc produces high-energy ultraviolet radiation which burns the eye's retina if viewed either directly or indirectly—and blindness can result from too-long exposure. Short doses of UV light give 'arc eye'—a painful, 'gritty' feeling with watering, which comes on a few hours after exposure. Bystanders can get it too, either from reflected light or by looking alongside the arc. UV light will also sunburn exposed skin. It's good practice to paint the work area in a dark matt colour. When working outside a dedicated welding bay, be extra careful around helpers, children and curious livestock.

'EW' on a filter glass means it's suitable for electric welding. The higher the code number, the denser the filter on a scale from 8 to 14. Use EW 10 or 11 for most work; lower numbers only protect up to currents of 100A, while the highest numbers cut too much light at farm-size welding currents. Whatever you do, NEVER use gas welding goggles for arc work—they don't filter UV light at all. Protect arc filters with a clear plastic or glass screen, wiping this clean frequently during work (B). You can't see much through a fog!

1.3 Three alternative sets, each giving different periods of operation. The oil-cooled, copper-wound set (A) is designed for continuous use: it'll run as long as needed on full blast. The AC output model shown has both 50V and 80V output lines (bottom, left and right) giving flexibility over electrode type usable; these are open circuit voltages or OCVs (see page 25 for explanation).

The fan-cooled set (B) has vents for forced-air passage. It'll run until its windings get hot, then a thermal trip operates. After a short cool-down period it comes back on stream. This Murex example produces DC output which makes for a smoother-running arc, but it's more expensive than the AC alternative.

Very common on farms is the simple air-cooled set (C), which is quite capable of producing acceptable work. There's no forced-air cooling, which means a shorter operating period before automatic shut-down and a longer wait before restarting.

Duty cycle *should be checked before buying an air-cooled set. It's the amount of time within a ten-minute period that the unit can operate at the current quoted. Thus a duty cycle of 60% at 120A means the set will run for six minutes in ten at 120A; at lower currents it'll run for longer. The higher the duty cycle, the longer work can go on without the set shutting itself off to cool.*

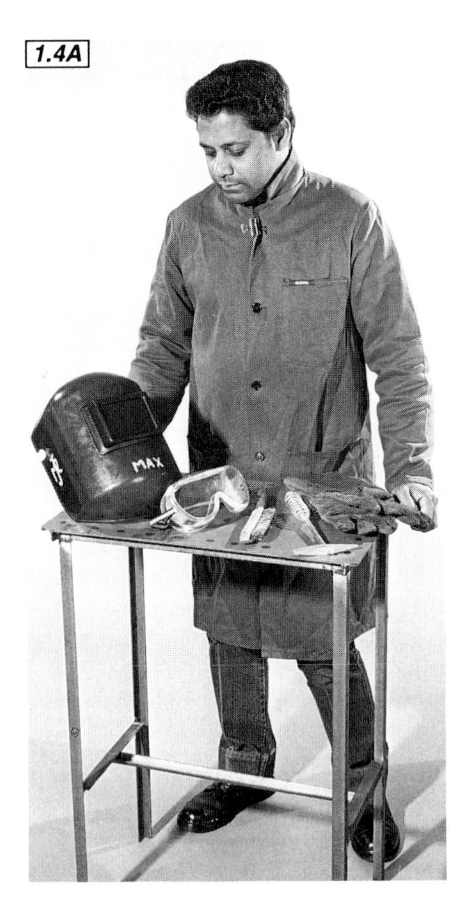

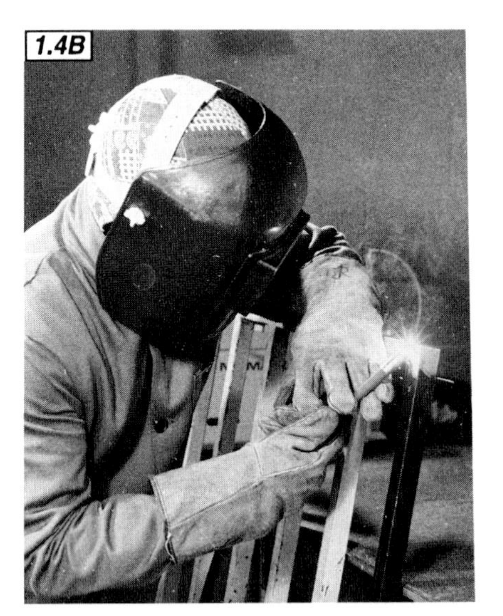

1.4 What the well-dressed welder is wearing (A). Leather boots for the feet, as canvas or rubber footwear offers no resistance to white-hot metal droplets. Cotton trousers without spark-catching turnups, and a non-synthetic overall or jacket done up at the collar to stop unwelcome sparks and slag. Synthetic materials may catch fire easily, so avoid them. Gauntlets are good, both to protect against cancer-causing ultraviolet light and to stop hot metal disappearing up your sleeve—neither is a good joke. Thick leather for the gloves means no burnt fingers when handling hot bits, either. The head is protected by a good-quality mask, along with a cap if appropriate (B). A flip-down welding helmet lets the operator use both hands when welding or tacking, but doesn't suit everyone. Throw away the lolly-on-a-stick mask that often comes with a cheap welding set—it won't be big enough to give full protection. Goggles (A) look after the eyes during slag chipping or grinding.

1.5 Electrode fume is not the best of stuff to breathe, and COSHH legislation insists on operator protection. Ideally, remove fume at source with a fixed or mobile purpose-made extractor (A). Where fume concentrations are relatively low, a ventilated welding visor (B) will probably suffice. It may be that no equipment at all is needed: if work is intermittent and good air movement can be achieved by opening doors/windows, fume will be diluted and moved on. Check with your local HSE Inspector. Whatever arrangement is used, always try to work out of the rising fume column (C).

Where possible, minimise fume problems—and maximise the chances of a good weld—by grinding away all paint and plating for 50mm on either side of the weld line. Never work in the blue-white fumes given off by overheated galvanised plating (watch out for coating on the inside of pipes!) or sickness will follow. Drinking milk may help if you are caught out. Be extra careful with anything that's been cadmium plated, for cadmium fume is a killer. Various paints and degreasants give off poisonous vapours when heated, so don't take any chances with fume: ventilate, filter and/or extract.

1.5A

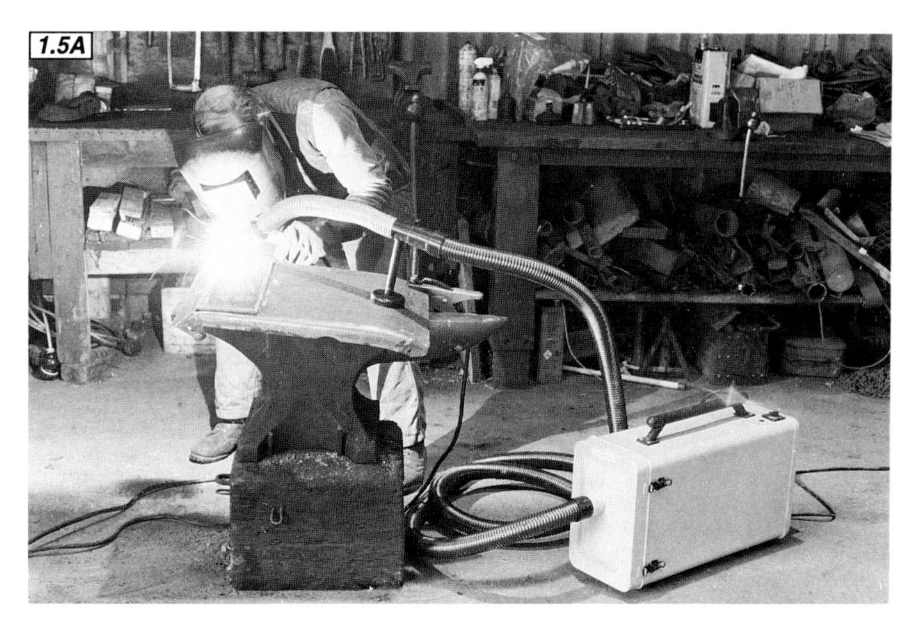

1.5B

1.5C

1.6

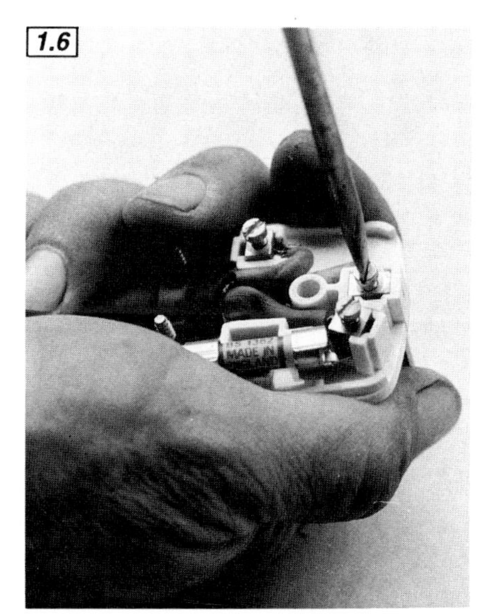

1.6 Mains supply plugs get warm in use, and repeated heating and cooling tend to loosen their terminal screws. What happens if a wire pulls out? At best the plant stops working. At worst you're dead through an unearthed set. Check for screw tightness every few weeks, but don't overdo the screwdriver as this can lead to localised overheating.

'Tingles' while changing electrodes are unpleasant and reflect the relatively high voltages present. They're usually worse in damp weather as current conduction to earth is better. Wear gloves or lay the new rod in your (hand-held) mask to sidestep them.

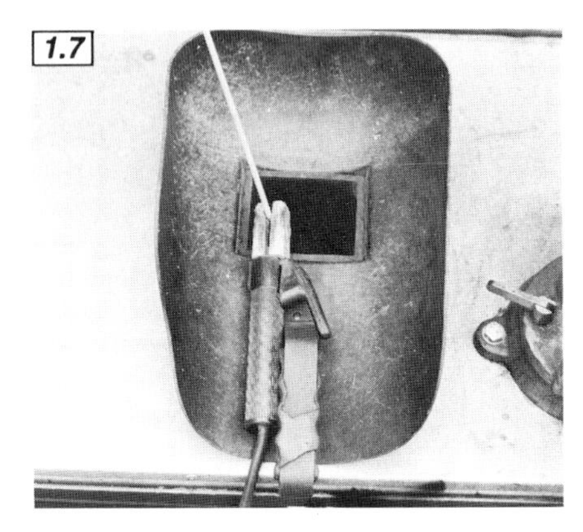

1.7 A good habit to cultivate: lay the electrode holder in your mask or on insulated material (wood is good) whenever it's not in use. Then you know where it is, and the chances of accidental arcing are greatly reduced. Take out the electrode when feasible. Give any form of gas cylinder a wide berth when handling the electrode holder, as accidental contact can be terminal. For this reason, never chain cutting/welding cylinders to a metal welding bench.

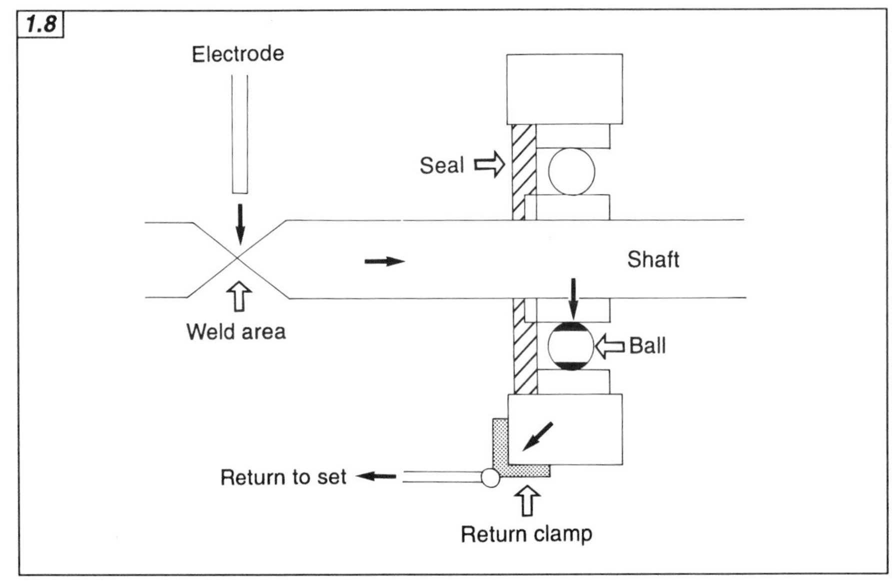

1.8 Watch out around shafts and bearings. Putting the return lead clamp on a bearing housing means that current goes through ball or roller contact patches on its way to the clamp (black arrows). Burning/flat spotting can result, turning the bearing to junk. Consider how fast and far heat can travel—seals are easily cooked by heat moving up a shaft.

1.9 How many potential nasties can you spot here? Answer at foot of page 6.

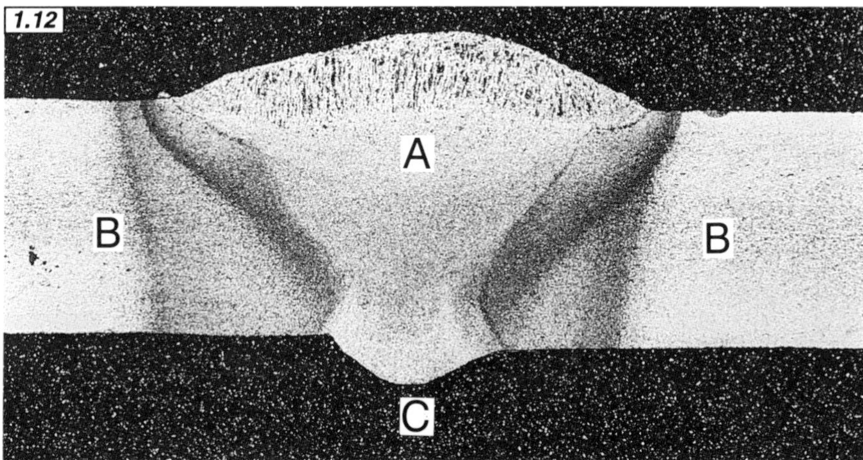

1.10 Early alternators could be damaged by the welding set's relatively high voltages, but modern versions have inbuilt protection. If in doubt, disconnect an alternator before welding on the vehicle or anything attached to it—better safe than having another bill to pay.

1.12 Sectioning a sound weld shows how weld metal (A) has fused with the two parent plates (B), and penetrated through them (C). Achieving these Terrible Twins—fusion and penetration—is the key to a strong, durable weld: fail to achieve either and strength will be lacking. Note also the way that weld metal is built up slightly at the top and bottom of the joint to give correct profile. A smooth blend between weld and parent metals avoids stress concentration, lessening the chance of failure.

Four runs (or passes) went into this joint. See how only the top weld retains a coarse grain structure. The rest have been refined and strengthened (normalised) by the heat of the following pass. Hence, several sound smaller welds are usually stronger than one big one.

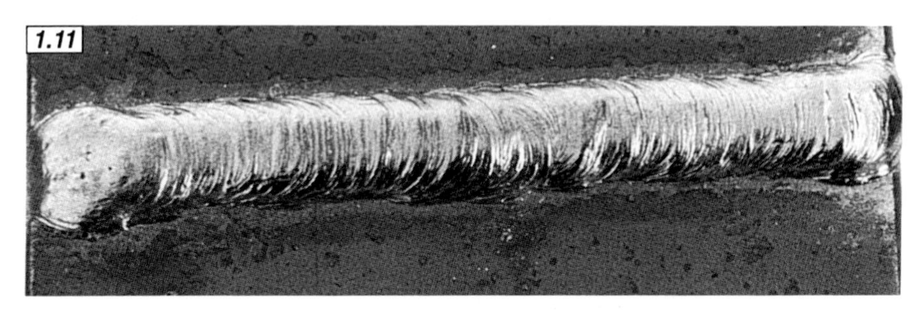

1.11 What you're aiming for—a good MMA weld. Note the even U-shaped ripples, good fusion at edges and lack of spatter. Good welds come from getting the four variables right—see page 8. For what's happened inside a good joint, see 1.12.

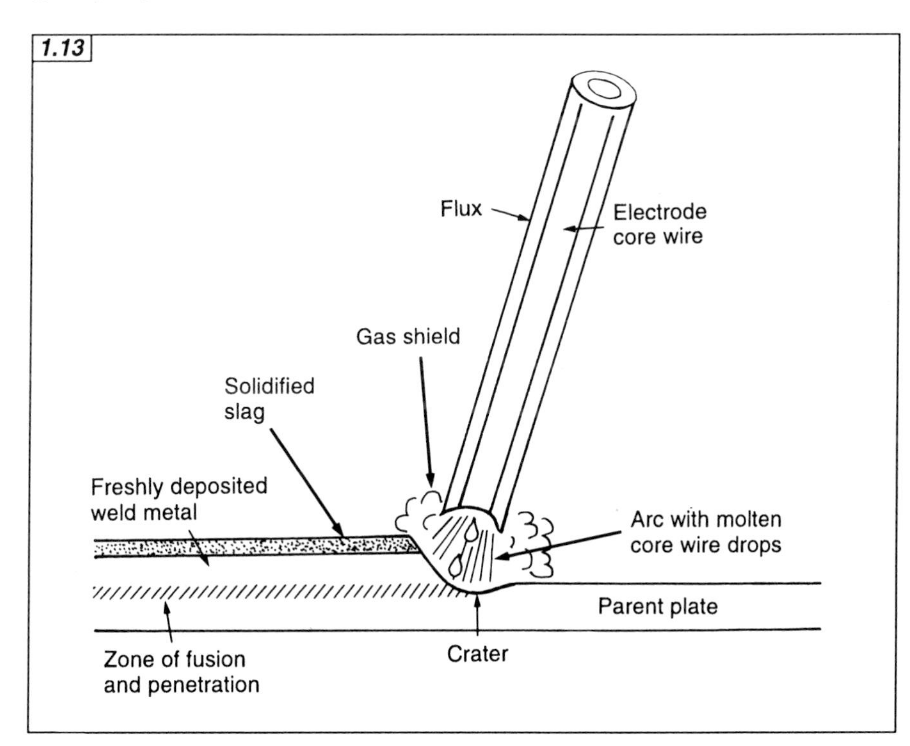

1.13 This is what goes on at the rod tip when current is correctly set (page 9). An arc spans the gap between the electrode and work, generating temperatures around 5,000°C. The outer rod coating burns, producing gas to shield the liquid weld pool from atmospheric contamination. Simultaneously, fluxing elements from the coating float weld impurities to the surface, where the mix solidifies as a crust of slag. Extra metal is freighted into the joint area as molten droplets of core wire, which are pulled across the arc by magnetic force and surface tension. That's just as well, or you couldn't weld up or downhill.

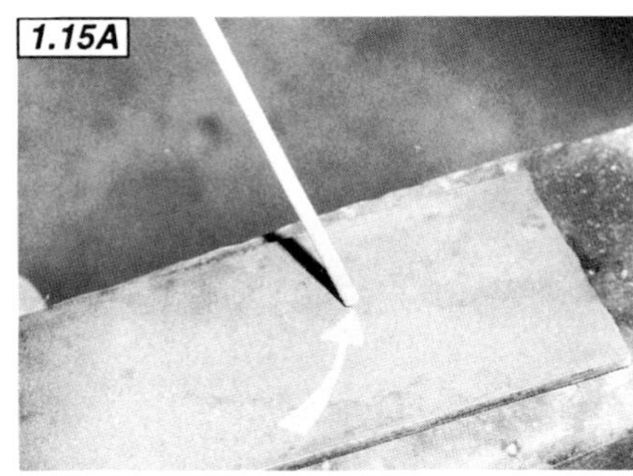

1.15 Getting an arc started is often frustrating for the beginner. The basic principle is simple: first scratch the rod tip to get it hot (current flows easier from a warm tip) and then quickly TAP AND LIFT *the rod on the work. Be positive with the scratching; use a wide sweep (A) that will generate showers of sparks. Then bring the rod quickly into position over the weld area, and in one smooth movement tap it down on the work and lift it a little (B). The arc will start on the tap and establish during the lift. If the rod sticks, break contact quickly by circling and pulling up. Figures 1.16 and 1.17 suggest ways round poor starting.*

1.14 Operator relaxation is central to good control. There are two schools of thought on this: either take every chance to rest and support the body on bench or machine, leaving only the welding arm free to move, or learn freestanding control in a legs-spread position, as here. This way, the operator is happy to work when there's nothing to rest on.

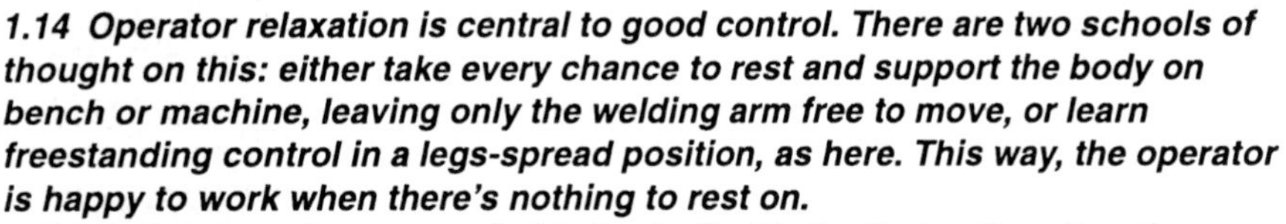

Occasional welders are probably best off with the first option. Practice leaning on the bench or machine. Take the welding lead's weight off the arm by looping it over the shoulders, pinching it between hip and bench or hitching it round the vice or toolbox—It goes without saying that the lead's insulation must be sound . A flip-down visor leaves one hand free for body support or (preferably) for steadying the welding arm. Whichever strategy you choose, don't grip the electrode holder tightly—it's a sure way to produce wobbles at the rod tip. Consciously relax your grip during a weld and see how control improves!

Answer to 1.9. Hazards on bench: Battery on charge (sparks mean possible explosion as hydrogen gas is given off during charging); rubber gloves and cloth at back (likely to catch fire); electrode holder not in mask and rod touching work (accidental arcing); cans for petrol, penetrating oil and engine lubricant alongside the welding area; an upended chainsaw (fuel leaks). On floor: Open waste container, cleaning rag box and grease bucket (fire hazards); inflammable degreasant; oil-soaked sawdust.

1.16A

1.16B

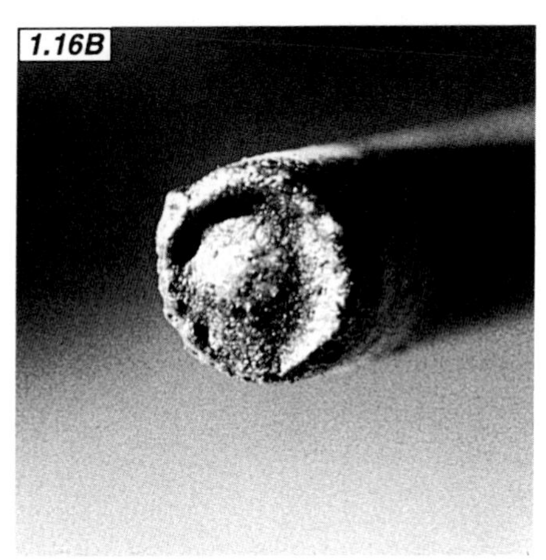

1.16C

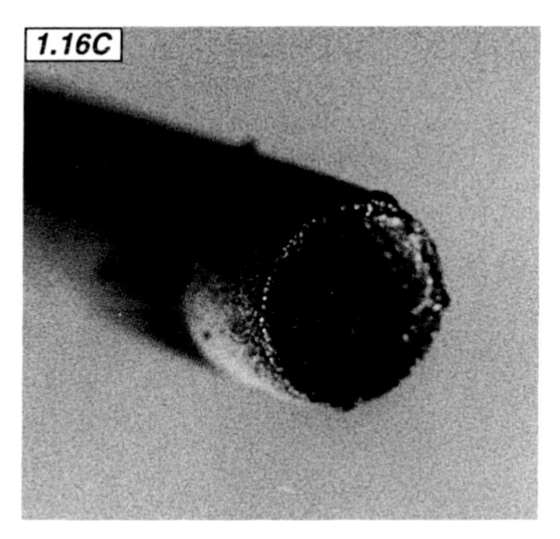

1.16 If sticking is a problem, look at the rod tip first. Is it like A or B? Condition A comes from heavy-handedness with a sticking rod, or damp flux. To rectify, scratch the rod hard on some scrap until the bare core wire is melted away. Condition B comes from a carelessly broken arc at the end of a previous weld; the slag has sealed over the core wire. Scratch and/or tap it hard (again on some scrap) to break the seal. Condition C is good—bare core wire waits inside a cup of hard flux, ready to establish an arc.

1.17(1)

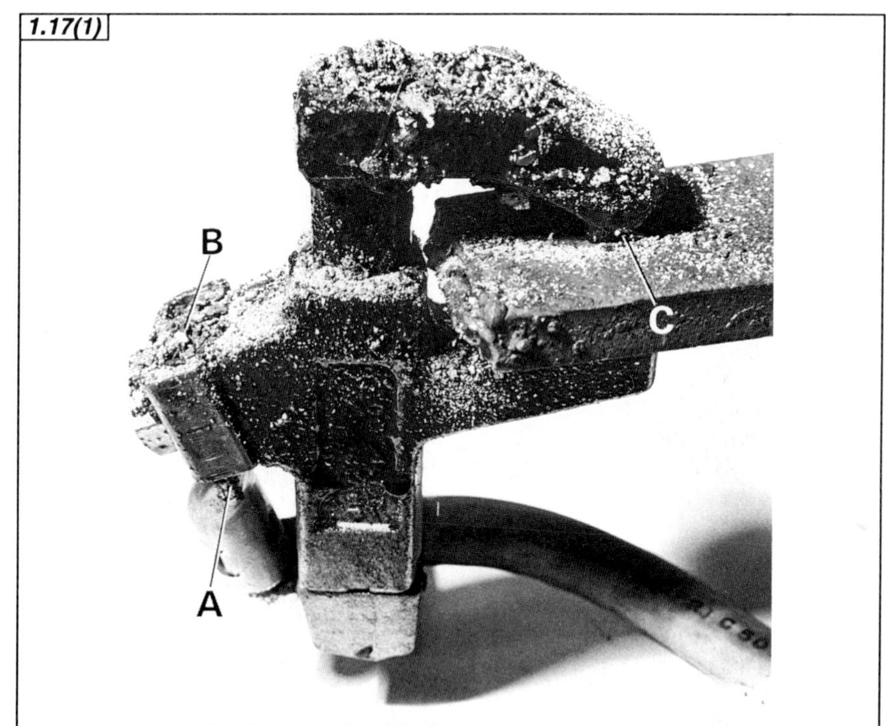

1.17(1) If the rod tip is healthy, look elsewhere for the cause of poor starting.
Good current flow is vital, but a return lead clamp like this chokes it. The contact point between clamp and work must be clean (C), so grind a bright area on the work or welding bench and attach your clamp there. If the return lead can't attach directly to the work, make certain there's good contact between the work and welding bench. For maximum current flow, all wiring must be in good order with no current-blocking 'bird's nests' of frayed connections (A & B). New return lead clamps are cheap, so throw out anything that's substandard and look after the new one. Don't use it as a striking-up area; a piece of scrap is a much better bet. Rod angle can hinder good starting; beginners often are tentative, laying (rather than tapping) the rod to the plate at too-shallow an angle. Come down firmly at a near-vertical angle—see 1.15B.

1.17(2). If the arc still won't start, check that there's enough current dialled in on the set. But how much is enough? See the rod pack for suggested setting range, and aim towards the top of that for easy starting. In this case it's 90A-140A for a 3.25mm rod. The pack also shows the minimum open circuit voltage (OCV) needed; here it's 50V. If the set can't deliver, striking will be near-impossible.

1.17(2)

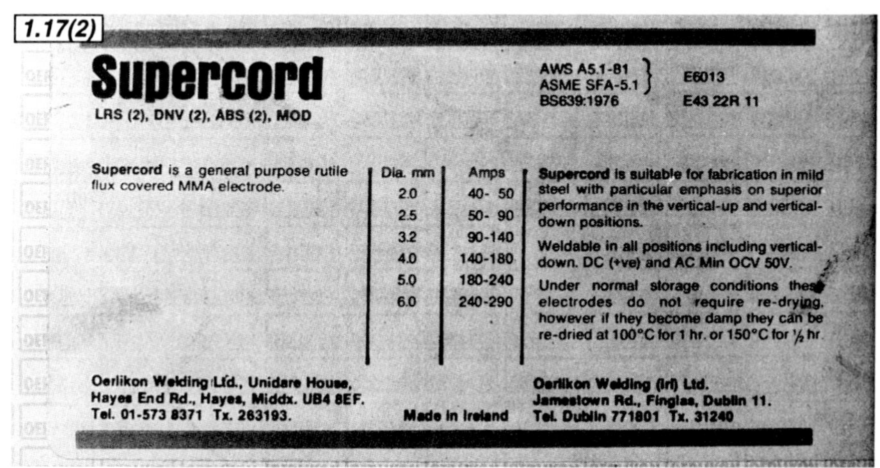

Finding the Rod for the Job

Which rod should you use for a given job? The variables are diameter, core wire make-up and flux type.

Take diameter first. The narrower the rod, the smaller the arc it produces. Thus thin section steels need skinny rods if arc heat isn't to burn straight through them. Thick sections need more heat for fusion, and that calls for a bigger arc—so use a bigger rod.

Practise first with 2.5mm diameter rods, as these are shortish and relatively easy to control. They'll do for welding sheet up to 3mm and can be used for preliminary runs in much thicker stuff. As plate thickness or joint area to be filled grows, move up to rod sizes 3.25mm and 4mm. These three diameters cover the majority of farm work.

The electrode's core wire must match the parent metal if good fusion is to result. Thus mild steel work needs a mild steel electrode, while fancier stuff requires fancier rods. Note that it's a waste of money using a tough, expensive alloy steel electrode to weld mild steel. If welds are failing, technique is probably at fault.

Flux covering is tied to electrode type and application—the user has no choice here. Most general-purpose mild steel rods have a flux based on the mineral rutile, spiced with other ingredients to improve arc stability, reduce moisture pick-up and generally make them more sociable to store and use.

Cellulosic coverings enhance penetration, and turn up on rods designed for specific purposes like vertical down work and pipe welding.

Low-hydrogen coatings appear on rods intended for use with high-strength alloy steels, possibly also for service at low temperatures. They are very sensitive to moisture content and must be baked before use.

Some electrodes have iron powder added to their coating. This adds to the volume of weld metal laid down when compared with ordinary rutile-covered electrodes.

So which of this lot do you go for? In practice, rod selection for general farm use is pretty straightforward. Find a good specialist supplier (not necessarily the local agricultural merchant!) and buy mild steel rods from a major maker like ESAB, Murex, Oerlikon or Eutectic. Minor brands can be good (and are usually produced by one of the big makers anyway), but are best experimented with over time. Starting with a big-name brand is always a good idea.

Stock up on sizes 2.5mm, 3.25mm and 4.00mm, buying a pack of 1.6mm 'sparklers' if sheet steel work is intended (these latter items can be expensive). Where vertical down welding is to be done, be certain that rod flux produces fast freezing slag—see page 16.

As soon as anything other than mild steel shows up, lift the phone for advice from the supplier or rod maker, quoting your set's OCV capability (page 2). A good outlet will give advice freely, and provide maker's lists of electrodes and their applications for future reference.

'Dissimilar steels' rods cover most agricultural eventualities. For example, Oerlikon's Inox DW will weld steels of varying carbon content, low alloy and spring steels and ferritic and austentitic stainless steels to themselves—and each other. And as they're relatively hard, most dissimilar steels rods can be used as a 'buffer' layer between mild steel and a hard-facing layer, reducing the risk of dilution which can cause the hard top coat to crack and flake away.

Buy only such rods as required, as they're expensive. Try to steer clear of low hydrogen electrodes, as they're a hassle; in non-critical work, dissimilar steels rods will probably cope anyway and are far less trouble to store and use. Cast iron is a law unto itself and needs dedicated rods and technique.

Similarly, building up worn parts or laying hard metal on new soil-engaging bits calls for specialised hard-facing consumables. Go to the supplier, tell him the application and he'll recommend something appropriate.

The cardinal rule with any work (particularly where materials other than mild steel are involved) is *'Be cautious'*. If in doubt over technique or rod selection—especially where the repair is safety-related—*have the job done by a specialist*. The alternative can be most costly in all ways.

Finally, a word on rod storage. Keep everything in the original packet, thereby retaining information on rod type, application and conditions for use. Ideally keep rods dry—an old fridge with its light switch modified so the bulb burns continually is good, and at least you'll know the answer to an age old question: what happens to the light when you close the door?

Good Technique: The Four Variables

Cracking the Four Variables is the key to a good job. Get them right and you'll have a sound weld. It's easy . . . at least in theory!

Have a quick think about what the Four Variables might be. The welder has three of them under his/her control during work: **rod angle, speed of travel** and **arc length**. The fourth is **current**.

Between them, these variables sum up the manual arc operation. Get every one right and a strong job results; get one or more wrong, and faults appear—lack of fusion, poor penetration, slag traps, undercut—all the things that make welds weak.

Of the four, perhaps current (Fig 1.18) has the edge in importance. If you're short here, the Terrible Twins—penetration and fusion—will be lacking. These are central to a sound weld (page 5).

The first rule of current selection is not to believe the graduations on the welding set! With time you'll learn to see, hear and feel how the rod is running; in short, to monitor what's happening at the rod tip and alter current accordingly. Changes in welding position usually mean changes in current too, and these will be covered later.

Until an experience base builds up, start by using the rod pack's information to set initial current level (Fig 1.22). Choose a setting towards the upper end of the recommended range. More current means more heat, and more heat generally means better fusion/penetration. As a spinoff, arc striking will be easier.

If the rod packet information is missing, allow 40 amps per mm rod diameter. On all but the highest quality work, a little too much current is better than not enough.

To get a feel for current setting, experiment on some scrap. Make short welds, starting from below the packet-recommended minimum and ending above maximum; see how striking, maintaining the arc and fusion all improve as current goes up.

But learn where to stop. As current goes over the recommended maximum, light and noise emissions get more violent, the rod may glow red-hot, spatter abounds and there is the risk of producing porosity. The weld surface will look 'burnt' and have coarse ripples; these may be pulled into Vee-shapes, as travel involuntarily speeds up in response to furious happenings at the rod tip. Ultimately weld quality suffers from too much current, and burn-through might be a problem. Don't overcook things.

The other variables are covered in Figs 1.19–1.22. Generally, first-time welders use too little current, work too fast with too-long an arc and tend to let rod angle drop until it's too shallow.

1.18 VARIABLE 1: If there's not enough current (heat), weld metal just heaps on top of the underlying plate (A), leaving fusion and penetration poor. Where current is too high (B) the weld bead is wide and flat and the final crater deep; excessive heat melts a wide track. Here the surface ripples are coarse and were pulled into Vee-shapes as the operator speeded up to counteract fast material melting. Beads of spatter were thrown from a Vesuvius-like weld pool.

Undercut (B ,1) may turn up where bead and plate meet. As the arc passes, it melts a wider crater than can readily be filled by rod metal (especially if the arc is too long), and valuable metal can be lost as spatter. Slag is trapped at the weld edge and by the coarse bead surface, making it hard to chip off (B, 2). Watch out for excessive undercut, as it provides a notch from which breakage begins.

When current is about right, weld metal has both fused and penetrated into the parent plate (C). That's just what's needed for a good weld, though you obviously can't see penetration here. The deposit is of consistent width and the surface ripples are U-shaped; there's no undercut and little spatter. For most jobs, a little too much current is better than not enough. But don't go overboard or you'll waste rod, create undercut, risk burn-through, cover the job in spatter and increase distortion by feeding in excess heat.

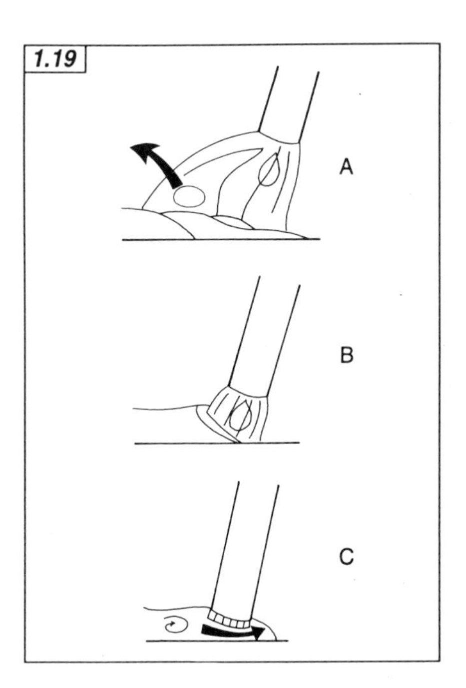

1.19 VARIABLE 2: Arc length. A rough guide keeps the arc just shorter than the core wire diameter, i.e. about 2mm for a 2.5mm rod. But it's by no means a fixed quantity. Experiment is again the best guide—you'll know when length is about right, for arc noise hardens from a soft, fluttery sound (arc too long, A) to a crisp crackle as the right length is found (B). A good-length arc sounds exactly like the sizzle of a fried breakfast!

Look hard at the rod tip during work. When arc length is right you'll see a clear, bright area under it, bordered by a semi-circular wave of molten slag behind. When the arc's too short, the rod dips in the molten slag and everything goes quiet (C). Arc light dims and slag wants to run round in front of the tip—a perfect scenario for producing slag traps (page 18).

As operator control grows, arc length can be used to control slag. Watch the receding slag; if you see a violent swirl starting to develop just behind the rod tip, a trap is imminent. Lengthening the arc a little usually saves the day. Be careful though: too long an arc will itself produce a trap as the weld pool isn't properly filled, giving space for slag. It's a common problem in fillet joints. Experimenting with arc length (while keeping other variables right!) is the best way to find out what's happening.

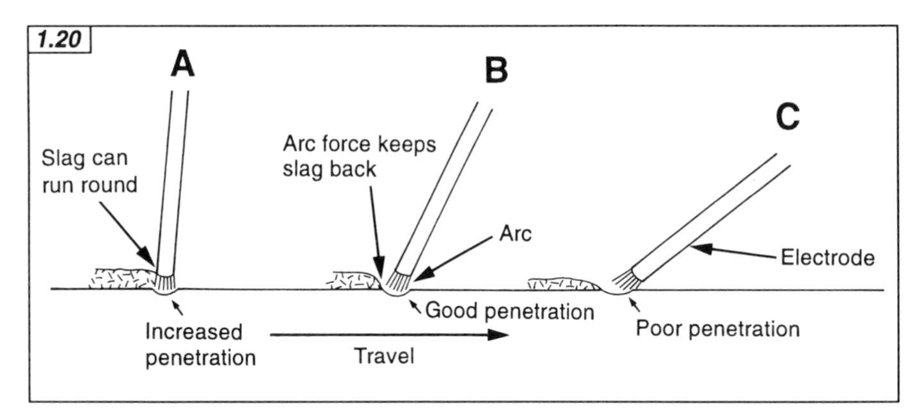

1.20 VARIABLE 3: Rod angle is important to fusion and penetration. With too shallow an angle, the arc's force blows weld metal away, giving poor results (A). But go too steep and there's a chance that molten slag will run round the rod tip and interfere with the arc, causing aero-chocolate slag traps (C). Proper rod angle balances these possibilities (B). For work on the flat, an angle of 60–70° is the norm. Some rods tolerate operation outside these limits better than others. Experiment with rod angle, looking and listening as the angle is changed. Where provided, angled grooves in the electrode holder give a guide. If you use these and keep the holder's handle parallel with the work, rod angle won't be far out for downhand (flat) operation.

When welding along a joint, keep the electrode centred over it. Angling the rod one way or the other puts more heat into one plate—bad for balanced fusion. But deliberately angling the rod to control heat input can be useful, as when welding thin sections to thick ones—see page 15.

1.21 VARIABLE 4: Speed of travel. How fast is fast? Too fast (A) leaves no time for weld metal to be deposited, so the bead is stringy and very weak. Travelling too slowly (B) takes extra time, wastes rod, increases distortion by increasing heat input and leaves a large humped deposit. By contrast, moving along at about the right speed (C) leaves a bead roughly twice as wide as the electrode core wire, i.e. a 4–5mm deposit from a 2.5mm rod. Look closely at the top bead and you'll see that it necks down in places along its length; the speed of travel was not constant, producing a series of skips and jerks. The surface ripples were pulled into Vee-shapes at each point and the weld changed thickness, so strength would vary accordingly.

The Joint Approach

Now on to joining metal. *Step one is to clean before you weld*; this should be written in letters of fire. Agriculture produces a good range of potential contaminants in the shape of rust, paint, oils and cow poop. These may not stop the arc, but all will compromise the weld's ultimate strength if not dealt with, no matter what your electrode salesman says!

Before proceeding, a slight philosophical deviation. To what extent is on-farm welding concerned with 'ultimate strength'?

Such a result comes only from thorough preparation, intelligent rod selection, appropriate plant setting and a high level of operator skill. Top-line work is the province of the trained man working to industrial standards; lesser mortals certainly don't have the training and practice, and may not have the basic hand/eye coordination of a fine welder.

But making a long-lasting, safe job is within most people's reach, and the cornerstone on which it's built is Good Preparation. So go ahead; invest time happily in removing crud and aligning parts, secure in the knowledge that your time is well spent.

What happens if preparation isn't thorough? Rust left in the weld area turns up as brittle, weak iron oxide particles in the weld itself. Thus for starters, grind off (or flame clean/wire brush away) all rust over and around the weld area.

Then there are paint, oil and animal products. All will vapourise under the arc, producing blowholes in the weld varying in size from the invisible to substantial pockmarks. On jobs where oil or other organic contamination has really penetrated (like an old oil-containing casting, perhaps) it'll probably be necessary to grind away a first, blowhole-riddled weld and redo. Otherwise, a run round with the angle grinder is normally enough to clear contaminants.

Aim to prepare an area some 50mm wide on either side of the joint. That way, potential fuming from nearby paint or plating is minimised. The joint will probably need a lick of paint afterwards, so

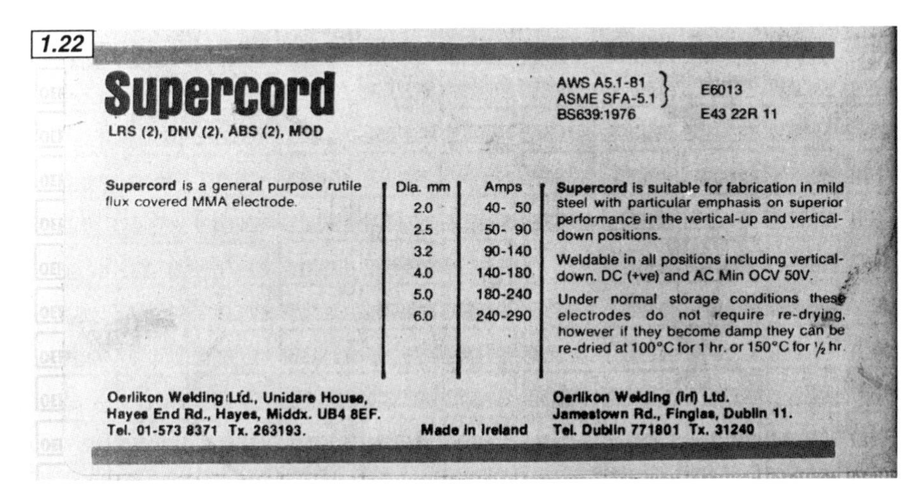

1.22

Supercord

LRS (2), DNV (2), ABS (2), MOD

AWS A5.1-81 / ASME SFA-5.1 — E6013
BS639:1976 — E43 22R 11

Supercord is a general purpose rutile flux covered MMA electrode.

Dia. mm	Amps
2.0	40- 50
2.5	50- 90
3.2	90-140
4.0	140-180
5.0	180-240
6.0	240-290

Supercord is suitable for fabrication in mild steel with particular emphasis on superior performance in the vertical-up and vertical-down positions.

Weldable in all positions including vertical-down. DC (+ve) and AC Min OCV 50V.

Under normal storage conditions these electrodes do not require re-drying, however if they become damp they can be re-dried at 100°C for 1 hr. or 150°C for ½ hr.

Oerlikon Welding Ltd., Unidare House, Hayes End Rd., Hayes, Middx. UB4 8EF. Tel. 01-573 8371 Tx. 263193.

Made in Ireland

Oerlikon Welding (Irl) Ltd. Jamestown Rd., Finglas, Dublin 11. Tel. Dublin 771801 Tx. 31240

1.22 Rods from reputable makers carry full information on the packet (don't buy if you can't see it): material(s) that the rod can weld, positions it can be used in, suggested welding current range and OCV. This last is open circuit voltage, and is the minimum figure your welding set must produce if the rod is to run readily (see MMA set, page 25). Equipment with more than one output voltage gives flexibility over rod types usable. If you have a DC set, see the pack for the rod's preferred polarity. Note: some rods are easier to use than others on vertical work, even though their packets may claim both are suitable. It's all in the slag, whose significance to vertical work is explored on page 16.

1.23 With the material cleaned, rod selected, starting current set and the Four Variables fixed firmly in mind, a good exercise involves laying down overlapping runs on a plate. Sounds boring? Maybe, but this basic technique will be used time and again when welding deep Vee-joints, building up worn parts and hardfacing. To do it, first lay down a single bead using a chalked guide line if needed.

Then add successive runs so that each half-overlaps its predecessor. To achieve this, use a 45° rod angle and fire the arc right into the place where the first run meets the plate. Adjust speed of travel so the new metal half-covers the old. Chip off, wire brush and inspect after each run. An alternative way of building up a pad of fresh metal is to lay down two single beads (spaced not more than three times rod diameter apart) and weave between them. Weaving and weave patterns are covered on page 16.

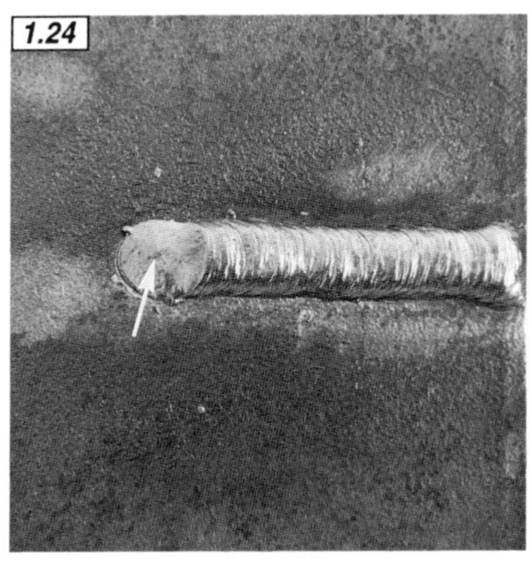

1.24

1.24 Sooner or later you'll run out of rod, so restarts are covered in the next four pictures. Burn off as much rod as possible, keeping an eye on the flux coat. If it's damaged or starts to fly away from the tip in beads, stop. Finish anyway when you're down to a 50mm stub. It's never worth trying to get to the end of a joint before the rod runs out—somehow you never make it, and just leave a poor weld anyway. As the old rod finishes, set things up for a good restart by breaking the arc cleanly; whisk the rod tip away to one side, leaving a shallow crater (arrow). Wire brush the area.

1.23

1.25 Get the new rod's tip warm by a short run on some scrap—not the welding return clamp. Quickly position the tip about 12mm forward of the crater (point X), tap it to start the arc, and immediately lift the rod. This does two things: gives enough light to see where you're going and minimises the metal laid down. Keeping the arc long, smartly move the rod tip back to the crater (dotted line). Close the arc down to welding length (hear its sound change), then make a very deliberate 'wall of death' passage around the crater to fuse old and new metal (solid line). Don't rush! Move slowly forward out of the crater, picking up normal welding speed as it's left behind.

1.25

1.26 Be slow and deliberate over restarts. Waving the rod tip around produces a mess (A, arrow). When done well, the restart will be hard to see and will not snag a fingernail run over the area (B, arrow).

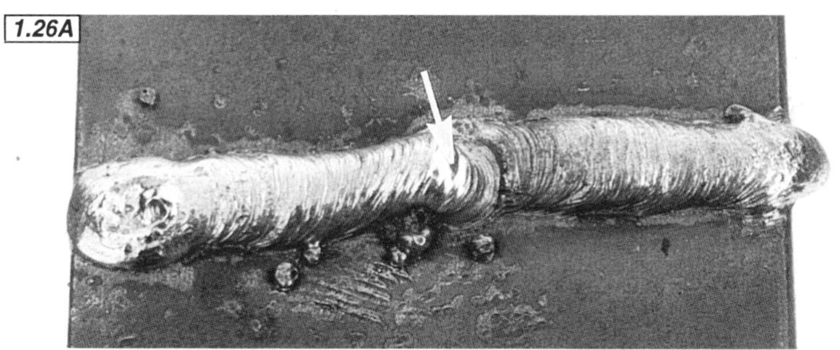
1.26A

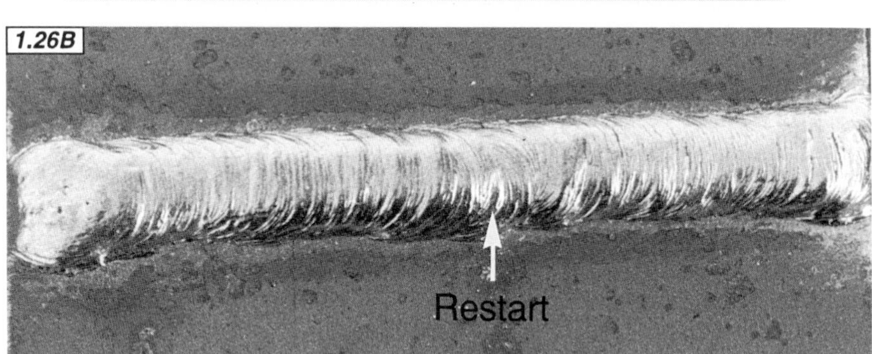

1.26B

1.27 There's another start type that gives trouble—one made at the beginning of a joint. A common fault leaves a 'snail trail' of weld and usually a slag trap for a short distance after the start (arrow); then a decent weld takes over. The problem is an initial shortage of heat, and hence lack of fusion. On starting, the rod and plates are relatively cold, so metal does not fuse readily; moving off too briskly then produces the fault. The trick is to start the arc just over (rather than on) the joint edge and to pause a while before moving off, watching to see that molten metal fuses with both sides. A small sideways weave usually helps the process along: try it and see.

1.27

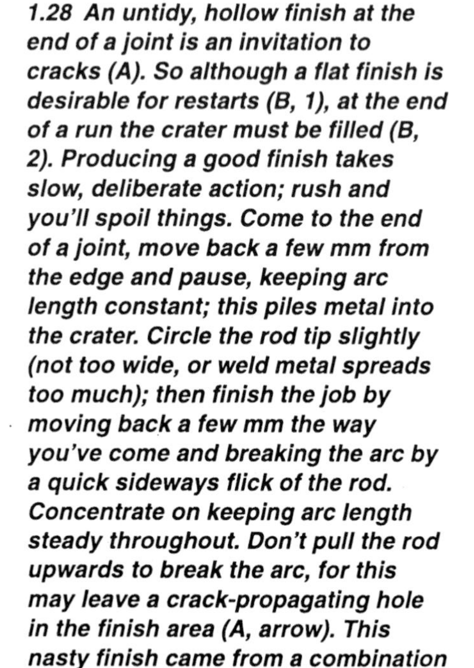
1.28 An untidy, hollow finish at the end of a joint is an invitation to cracks (A). So although a flat finish is desirable for restarts (B, 1), at the end of a run the crater must be filled (B, 2). Producing a good finish takes slow, deliberate action; rush and you'll spoil things. Come to the end of a joint, move back a few mm from the edge and pause, keeping arc length constant; this piles metal into the crater. Circle the rod tip slightly (not too wide, or weld metal spreads too much); then finish the job by moving back a few mm the way you've come and breaking the arc by a quick sideways flick of the rod. Concentrate on keeping arc length steady throughout. Don't pull the rod upwards to break the arc, for this may leave a crack-propagating hole in the finish area (A, arrow). This nasty finish came from a combination of poor technique and too-high current.

1.28A

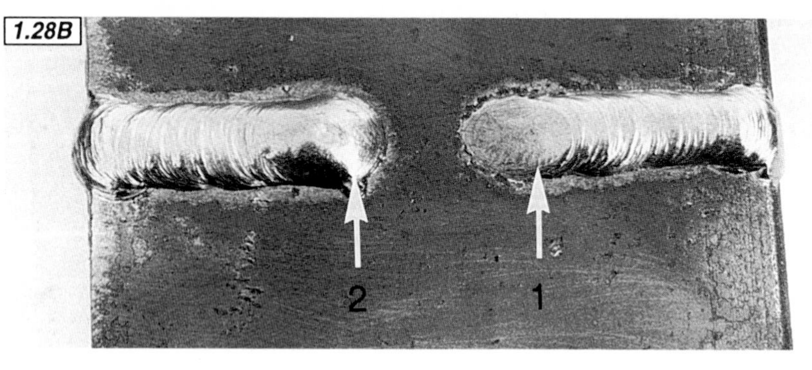

1.28B

generous preparation isn't wasted.

Metal platings have to be really watched. Recognise cadmium plating's dull yellow finish and grind all traces of it away, for heavy metal fumes are really nasty.

The zinc in galvanising is a problem, too. On heating it makes a mess of a weld, and the dense white fume produced does you no good at all. Where possible, use God's own ventilation—work outside in moving air. Inside, use a fume extractor and/or appropriate respirator.

Preparing to fire

After cleaning the surface, look to preparing the joint metal. It's hard to overstate the importance of this stage, but it's so often skimped. The object is to weld any joint to *full depth*, as only then can you hope to duplicate the original material's strength.

If you're talking butt joints (i.e. two plates laid side by side and joined by their edges), full depth work almost always involves some sort of edge preparation. Material thickness is the trigger: butts in up to 3mm mild steel can be welded without preparation, but something must be done with greater thicknesses.

(Note: blowing holes in the work gets increasingly likely at thicknesses below 3mm, so this can be regarded as a lower working limit for most MMA operators.)

Edge preparation lets the electrode reach the bottom of a joint. Just leaving a gap between otherwise-unprepared edges does help, but not much. For all its noise and fury, an arc does not penetrate well.

Check this for yourself by a welding series of butt joints in 5–6mm plate, leaving an increasingly large gap each time. Weld one side of the joint only, and then break it open in the vice and look for penetration.

You'll probably conclude that the arc needs help to penetrate, and the best way to give this is to bevel the plate edges with a grinder or gas cutter (Fig 1.29).

(*text continues on page 16*)

1.29 To get to the bottom of a butt joint in material over 3mm thick, first get rid of the edges with a grinder or gas cutter (dotted lines). This applies to both repair and fabrication work! It's generally reckoned that a 60° included angle gives the best compromise between accessibility and void size, but give yourself a little more room by generous preparation. This makes life easier, particularly with the first (or 'root') run which can be a slag trap generator if the Vee is too tight. Technique faults (like insufficient current, travelling too fast, wrong rod angle or too short or too long an arc) are magnified when welding in a Vee.

Go for good fusion and penetration on the root run (1). If the Vee is at all tight, use a little more current than usual. And if the situation allows, help penetration by leaving a gap between plate bottom edges (maximum width equal to electrode thickness).

Fire in run number one using a small electrode; 2.5mm is usually fine. Using too thick a rod at this stage defeats penetration, as a bulky rod can't reach into the Vee bottom as well as a thin one. Having said that, a thicker rod's extra metal and current can help stop slag traps—try it and see, but check penetration.

1.30 Further runs are then made to fill the Vee, using the overlap-producing technique covered on page 11.

Place each weld so it either overlaps or stays well clear of its neighbour (A), as the last thing you want is a narrow, slag-trapping gap between adjacent runs (B). As you move up the joint weld width grows, suggesting a change to thicker rods. Swap rods rather than going more slowly with a small rod to lay down more metal, or slag traps can result where there's not enough filler metal. After each run, chip off all slag and exercise the wire brush. Good 'weld hygiene' leaves no bits of slag behind to form weakening inclusions in the next run.

Where appropriate, finish off the joint with a weaved capping run (Fig 1.37A). This blends in the weld's edges and 'normalises' preceding passes (page 5). If you didn't manage to achieve full penetration from the first pass, turn the work over, grind the weld line back to clean metal and seal it with a single run. That is, if the work can be turned over.

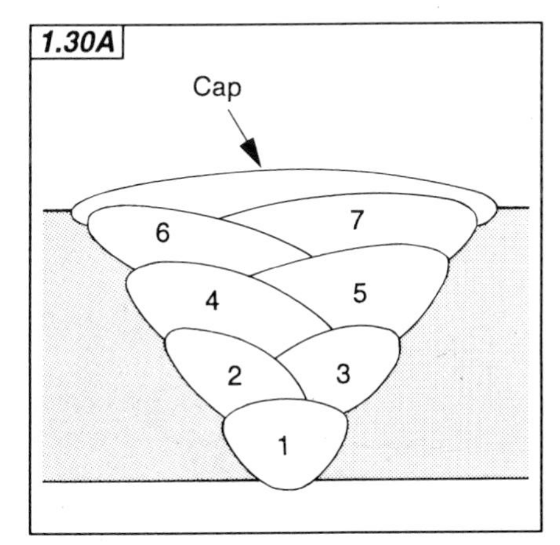

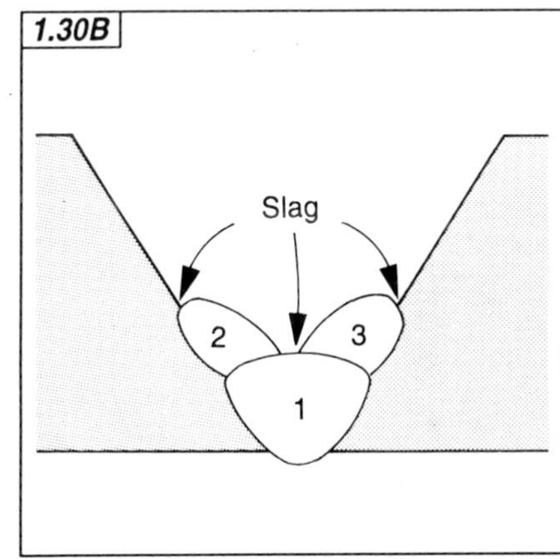

1.31 *Tacks need to be good-quality 'mini-welds'—not blobs—if they are to withstand contraction stresses. A meaty tack allows more latitude for correcting alignment (i.e. hitting with large hammer) than a blob, so good thick-material tacks are 10–12mm long and are fused properly into the joint material.*

Where possible, tack on the reverse side of a joint to leave the weld area uncluttered and to help minimise distortion. If this can't be done and you're after uniform surface appearance, grind away the tack's bulk before welding the joint.

Space tacks at 100–125mm intervals on long work. It's a good idea to:

(a) Save part-used 2.5mm rods for tacking. A short rod is easier to control, and its small diameter means good penetration in a Vee.

(b) Use more current than normal for an easy start and quick fusion. Do this by either shifting down one rod size (e.g. from 3.25mm to 2.5mm) but leaving current setting unchanged, or keeping a 2.5mm rod and lifting current by 15–20%.

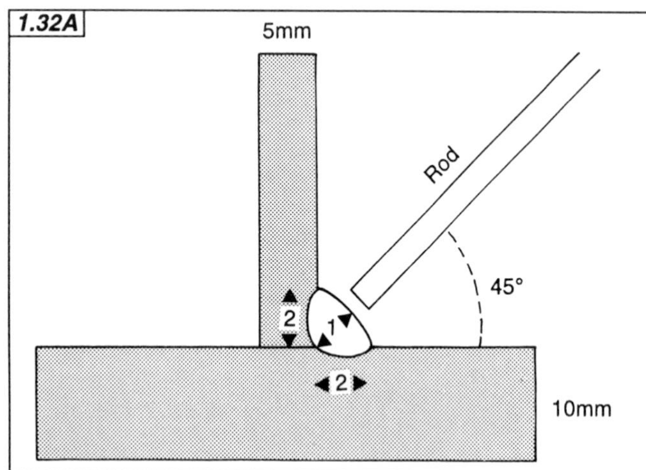

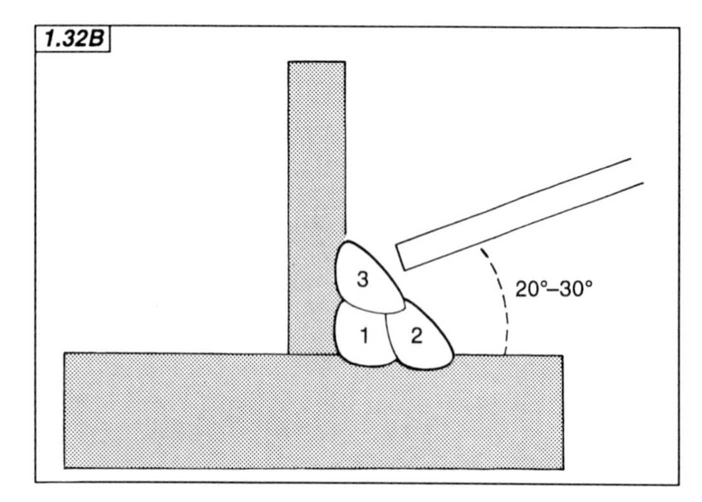

1.32 *To make a strong T joint, the weld's throat (1, A) should be equal to the thinnest plate. So, when joining 5mm to 10mm plate, throat depth should be at least 5mm. Equal leg lengths (2, A) also give balanced strength—achieve them by bisecting the joint with the rod (45° angle, A).*

Good root run fusion comes from a mix of careful plate alignment and steady work with a small diameter rod. On starting, let weld metal fully fill the joint before moving off, keeping speed down if necessary to ensure this. Keep travel speed steady and arc length short to ensure the joint is filled and free from slag traps.

Thick plates will need more than one run to get the required weld depth. In a three-run joint, sequence passes 1–2–3 as shown (B and C); see how run 2 gives the final pass a 'shelf' to sit on. Keep metal well up on the vertical plate during pass 3 by using a shallower rod angle. If appropriate, go up one rod diameter for runs 2 and 3.

If slag traps appear, first check the current—too little is a prime cause. T joints need more heat than butts, as the bottom plate is a big heat sink. Use 10–15% more current than with the equivalent butt joint, but don't overdo things; check the vertical plate for undercut at the weld/plate border—any notching here weakens the material.

If current is good but traps still appear, look at speed of travel. Going too fast means the joint can't fill, promoting slag traps and the likelihood of undercut. Arc length is also critical. If it's too long, metal goes to the joint walls (rather than the root); if it's too short, slag can run round the rod tip and interfere with the arc.

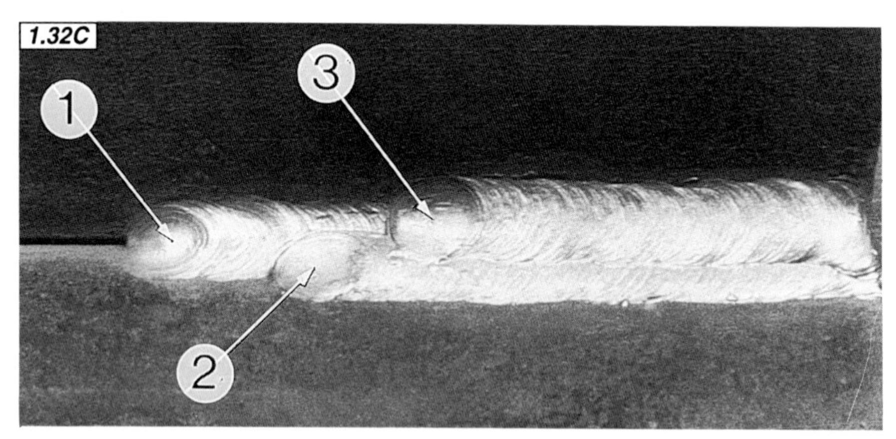

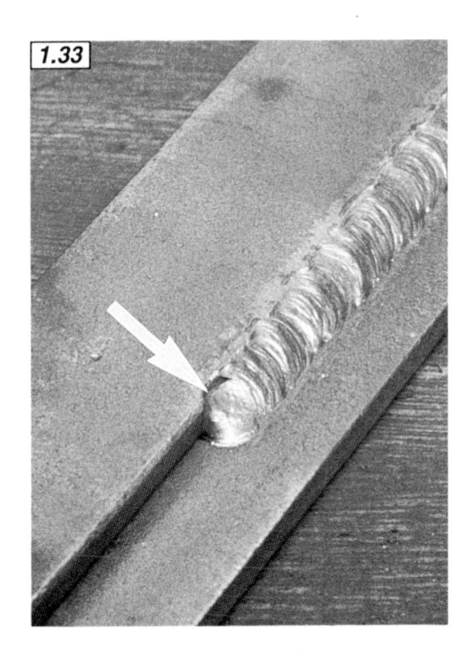

1.33 Lap joints are relatively easy work. Tack on the non-welded side where possible, and truck along at a speed such that the upper plate's top edge just melts into the weld pool (arrow). If the joint looks 'hollow' or won't fill properly, try the next-larger rod. As with T fillets, slag traps come from too little current, too fast travel or too long/short an arc.

1.35 Welding thin sheet to thick plate is not as hard as it might seem—so long as the two are kept in close contact. The picture shows 1.6mm sheet lap-jointed to 6mm plate. First the corners were tacked (one occasion where a quick blob-tack is called for), then each side was welded. The arc was aimed so that most of its heat went into the thick underplate, leaving just enough warmth to fuse the thin plate's edge (arrow). Watch the thin plate during operations; if it starts to buckle, it'll burn back quickly. Stop, dress it back down and then continue.

1.34 Open corner joints (90°) are equally straightforward, given good preparation. Set and tack plates so lower edges are parallel, leaving a small penetration gap in material over 3mm thick. Take care over set-up, as this is half the battle! Travel along the joint at a speed that sees both outer plate edges melting into the weld pool (arrows). Use multi-runs to fill the Vee in thick material, sequencing passes as Fig 1.30.

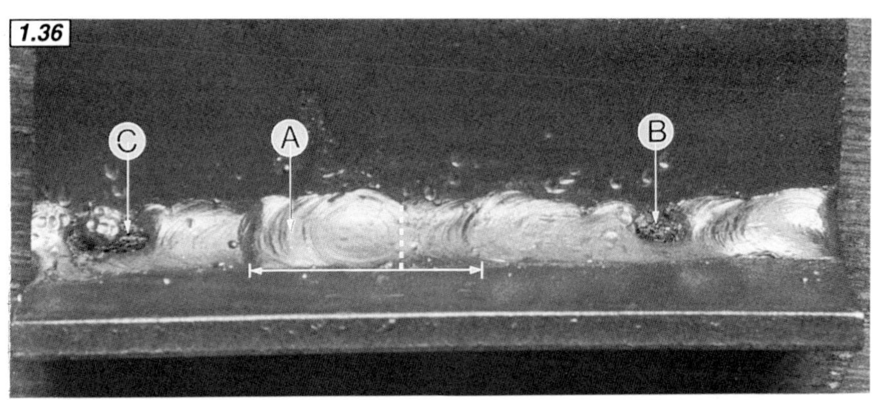

1.36 Slag traps happen to everybody! Good preparation, sound basic technique and lots of practice minimise their appearance. But what can you do when they rear their ugly little heads?

There are two options. Either grind out the joint line and reweld (not always possible) or burn out each trap using welding heat (not always 100% successful). Where possible, go for option one.

But the latter approach will do in a crisis. First hook out all get-at-able slag, and then reweld to float any remaining slag to the surface and fuse fresh metal into the just-vacated void. Violent measures are called for! Use a small electrode/big current combination to get mega-heat into a small area, say 130A on a 2.5mm rod. A common trick is to swap rods for the next size down from the one you're using, but leave the current setting where it is. Strike up just short of the trap, get the arc running nicely and then move slowly over the suspect area, watching to see that the arc digs out all the crud beneath it and that the void is filled with fresh weld metal. In the picture, trap A has been given the treatment, while horrors B and C wait their turn.

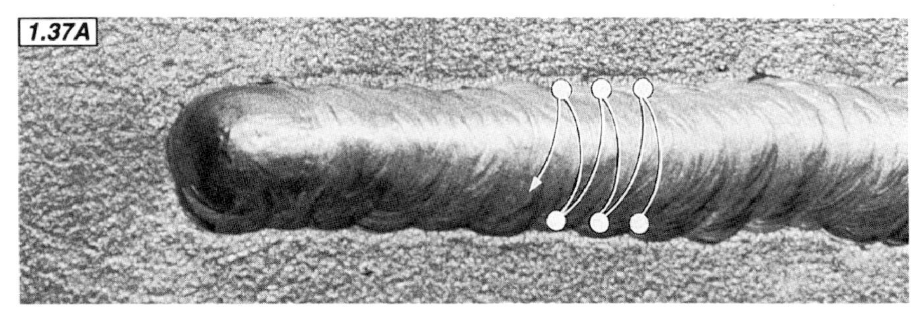

1.37 Weaves come in many forms. All have one of two aims—either to spread metal during surfacing, run-finish or wide-gap bridging operations, or to provide support for fresh metal during vertical upward work. Whatever the purpose, weaving is best kept simple—first because that's easier, and second to lessen the chances of trapping slag in the weld. Either a crescent weave (A) or zig-zag pattern serves most purposes other than vertical work.

How wide to weave? A rough rule of thumb suggests no more than three times electrode diameter; otherwise the weld's edges cool too much and slag traps and/or 'cold shuts' happen. In a cold shut, fresh metal lies over (rather than fuses with) the just-deposited metal. The weaver's maxim is 'Take care of the edges, and the middle looks after itself', which means making a deliberate pause at each side to let metal flow, fuse and build before moving smartly across the weld face (circles, A). As with restarts, make only very deliberate rod tip movements—waving the rod about invites slag traps. Work tight into the face of the weld, not travelling too far forward for each traverse. The technique of infill weaving between two beads can be useful where a conventional weave would be overstretched (B). Weaves for vertical work are covered next.

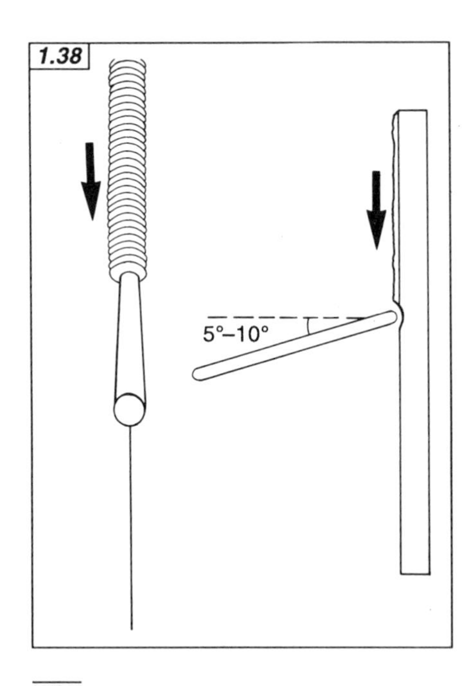

1.38 When welding vertically down, start with the right type of rod and with the current set towards the top of the recommended range. Centre the rod over the joint to put heat equally into both plates (left). Angle the rod just below the horizontal (right), and keep arc length shortish and travel at a speed that keeps the rod tip just below the decending slag curtain. The technique's inherent disadvantage of poor penetration can be turned into an advantage when trying to bridge large gaps or weld thin material.

Weaving

A useful technique to spread both heat and metal, weaving is used routinely in normal work when welding vertically up and for bridging large gaps. The important thing is to follow a definite pattern and work very deliberately. Figs 1.37A and B expand on technique.

The violent verticals

Vertical welding brings some operators out into a cold sweat, but it shouldn't. There are three options: move the work so it's welded on the flat (the Chicken's Choice, and not always possible), weld vertically downward or weld vertically upward.

Vertical down work is the commonly taken option, as most people find this easiest. But penetration—and hence strength—is always poor. Working vertically down is really only good for thin sheet or for filling awkward gaps where burn-through is otherwise a problem. *Never rely on it where strength is critical.*

Vertical down—the technique

Success hangs on having the right rod and enough current. Fast-freezing slag is called for; this stays up above the arc rather than running round it and interfering. Oerlikon's Supercord is a good choice, as is ESAB's OK Mildtrode. Use more current than you would when working on the flat, for arc force helps keep the molten pool clear of slag. Centre the rod over the joint line (Fig 1.38, left). Travel just fast enough to stay ahead of the descending slag curtain, with the rod angled just below horizontal (Fig 1.38, right). Keep the rod tip out of the slag, but don't let the arc get too long or weld metal won't find its way to the pool.

Vertical up

Working vertically upward produces deep penetration along with fear and loathing in equal measures.

Control of heat is everything. You can see the arc 'dig' as molten metal drains away from the crater—which is where the penetration comes from—but heat input must be closely controlled if the process isn't to get out of hand. If it does, the result is a series of dangling blobs, much like melted candlewax (Fig 2.36, page 41). So keep current down—use 10–15% less than the equivalent job on the flat. If you go too low in current, the rod will stick and slag traps form in cool areas. It's a balancing act: experiment at the lower end of the rod's recommended range, settling on the lowest current compatible with slag-trap-free, easy-running work.

The first pass in a multi-run joint is made without weaving, except when it's necessary to control penetration in a butt joint. Wider subsequent runs rely on a weave to support molten metal.

Weaving creates a horizontal ledge, which is used as a step to hold fresh metal. It's like making each rung of a ladder as you climb. Follow the arrows in Figs 1.39A and B to see where the rod tip went.

1.39 Two examples of vertical up work. The first is a two-pass job with two different weave patterns (A). The first run used a triangular weave; while going upward, the rod was pushed into the root gap until a change in arc noise indicated full penetration. The 'keyhole' shape left at the penetration gap (arrow) confirms this. Pausing at the weave border (circles) filled in the edges.

The joint was finished using an easier H-pattern weave—see lower part of job. Again, pauses were made at the weave edges to allow them to fill with metal. There were problems with this run—see how the second weave wasn't wide enough to blend with the plate's outer edges, and there's some undercut; too much current and/or too brief pauses were the likely cause(s).

An alternative is to use a simple crescent weave (B). This good-quality single-pass run was the result. Success came from very deliberate rod tip movement, coupled with a pause at each edge to let the crater fill (circles). When doing it, imagine you're a slow-motion robot. Current was low, the rod held at or just below the horizontal and the arc length kept as short as current allowed.

If your set permits, use the higher voltage line in vertical up work to help with arc establishment and continuity. Relax, take the cable's weight on surrounding metalwork and do likewise with body weight. Don't grip the electrode holder tightly, or control is harder.

Vertical up work usually has a wrinkled, prune-like finish; it's the result of all those individual ladder rungs. One or more downward passes using a crescent weave will tidy things up, but remember to increase current and use an appropriate rod (page 16). Various weaves are given in (C). Number 1 is the pattern for work on the flat; nos 2–5 are alternatives for vertical up.

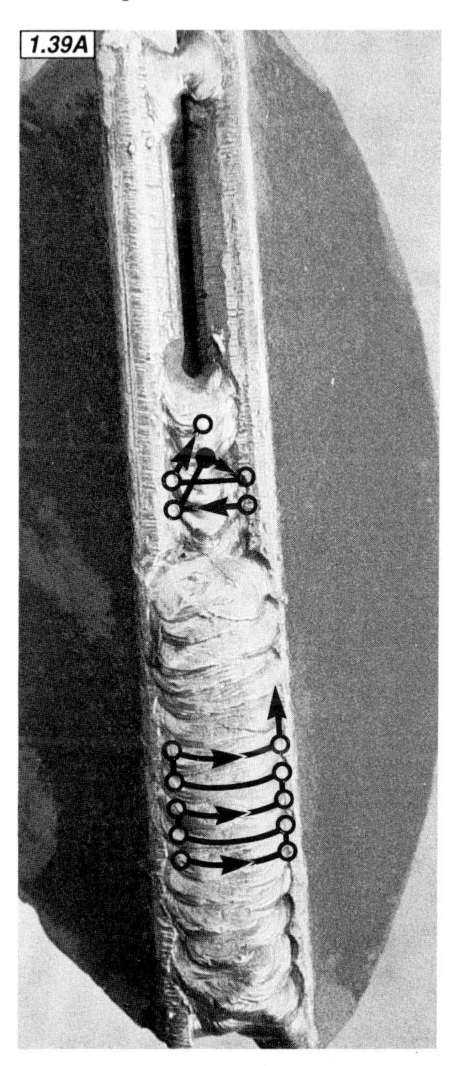

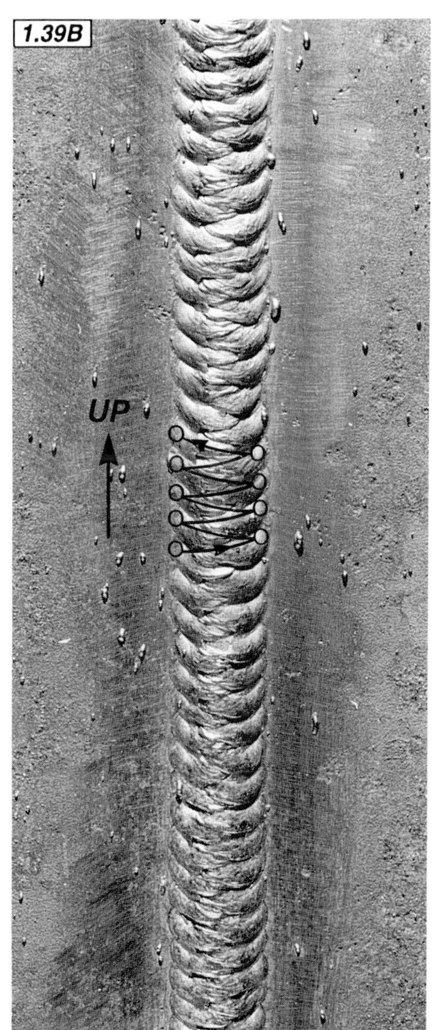

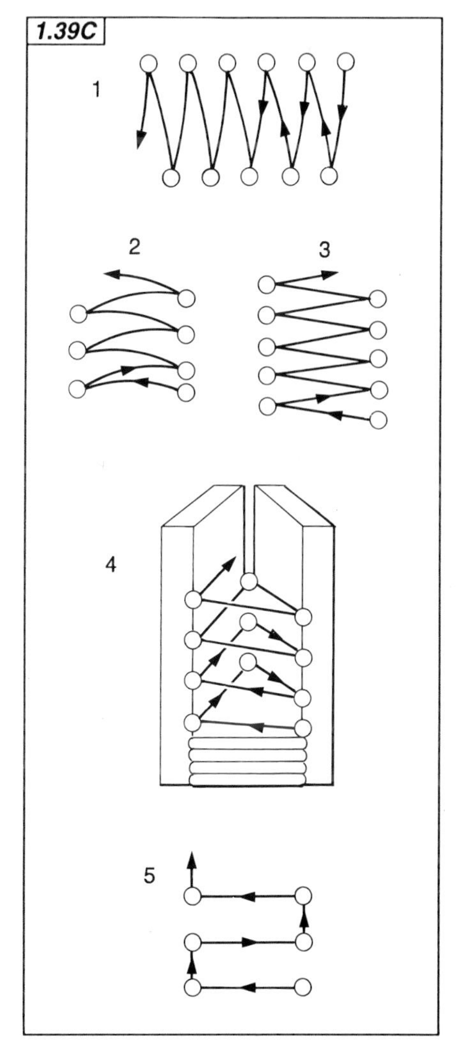

MMA Helpline

Everybody, beginner or older hand, makes mistakes. There's no shame in it. But if you want to improve, it's necessary to be able to spot faults, assess their seriousness and know where you went wrong. Figures 1.40–1.46 deal with common problems and ways round them.

Sorting out different materials is better done via the written word. All offcuts exhumed from under the bench will not necessarily be mild steel, so it's useful to recognise different steels and know which are readily weldable.

Accurate identification is a complex business. But main classes can be fairly readily spotted, given a file, a grinder and some basic ground rules.

Sorting the steels

Mild steel is the most user-friendly of metals. It doesn't harden when heated and cooled and is readily bent and welded. **Black mild steel** is the common stuff: it comes with rounded edges and retains its original mill scale coating.

Bright mild steel has square edges, a shiny appearance and is truer to size. It's made by cleaning and rolling black mild steel, leaving it stronger but less ductile.

Silver steel looks like bright steel, but is much harder, contains chromium but (oddly enough) no

1.40 Slag traps are the most obvious of problems, and the most corny advice is not to make them in the first place! Keep the Four Variables under control; use enough current and the appropriate rod angle, arc length and speed of travel. But the best-laid plans go astray, and the odd trap will turn up. The one above came from using too little heat; slag ran into the unfused centre (arrow). Dealing with traps was covered on page 15. Either dig or grind out slag, then reweld using a small rod/high current combination. Done properly, this remelts any remaining slag and fills the void.

1.41 Problems can hide under the shiny surface of a seemingly good weld. A combination of low current, a too-long arc and wrong rod angle left this T fillet with unequal length legs and a slag trap instead of root fusion (A). More care with current, angle and arc length produces a good joint (B). Here, a second weaved run had been added; note its coarse, grainy appearance. The underlying first run has been normalised by the heat of the second, refining its grain size and improving strength. The small hole at the root looks like a problem, but is an acceptable result of the vertical plate's original rounded profile. Lay a ruler along its right-hand edge to see this.

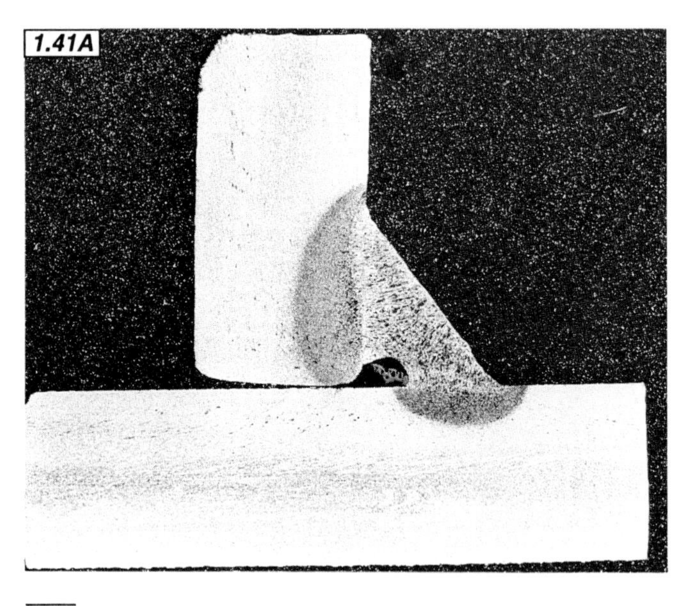

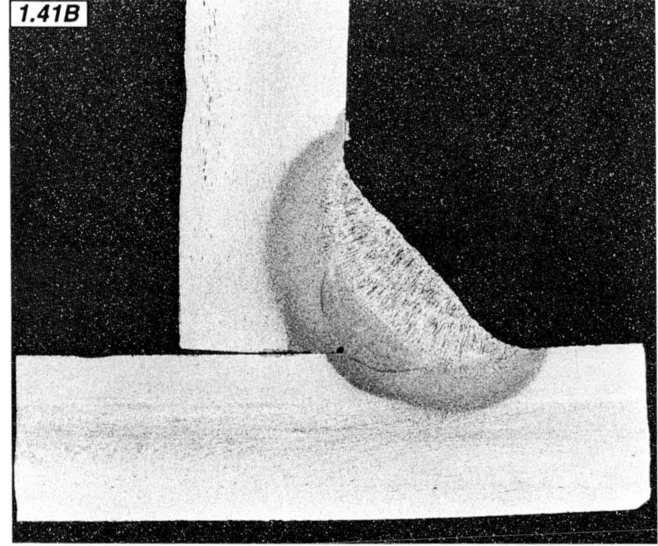

1.42 Two faults in one here. There's incomplete penetration (despite a penetration gap provided before welding) and a single, circular slag inclusion at eight o'clock. A smaller rod, run at a reasonable current, would have increased the chances of penetration and given enough heat to prevent trap formation. Just winding up the current on a big rod wouldn't solve the problem, for rod diameter is important where space is restricted. A big rod can't reach into a tight space.

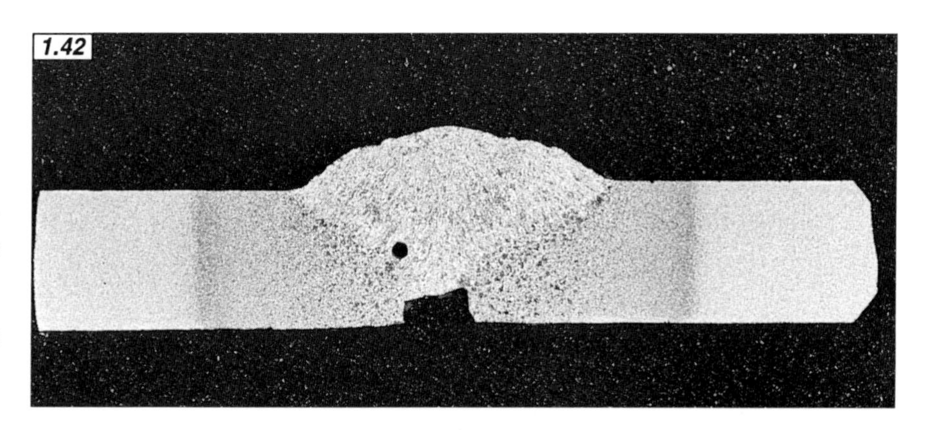

1.43 Poor preparation sank all hope for this outside corner joint. The plate edges were overlapped rather than kept parallel—see dotted line. This kept arc heat from the joint root and allowed slag to flow in instead. Aligning the edges so there's no overlap and (if appropriate) leaving a small penetration gap is the way round such problems.

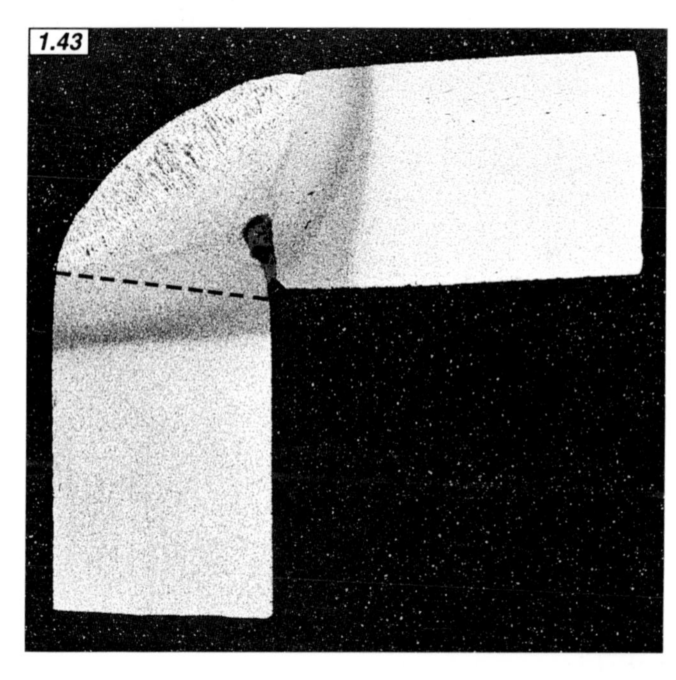

1.44 If too big a gap is left or too much current used, excessive penetration results (A, arrow). Weld metal from the first of two runs has oozed right through the joint. There's a good chance that it's weak, for sudden section change at inner corners means stress concentration. Uncontrollable burn-through was just a step away, too. Good penetration (B) fuses both lower plate edges without wasting metal.

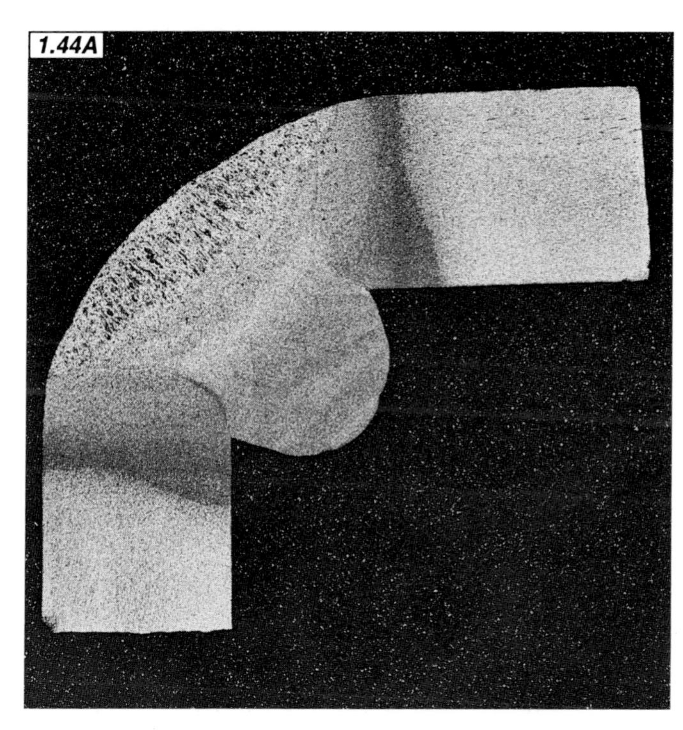

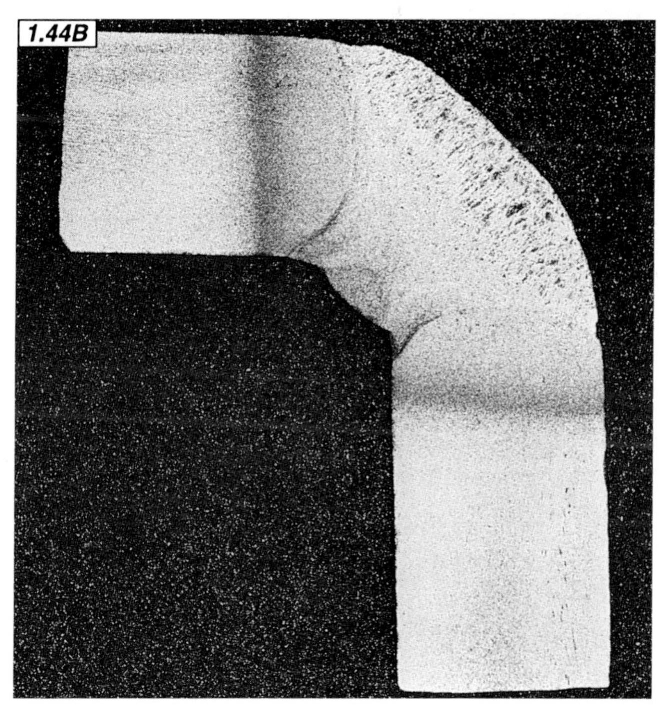

1.45 Nothing much wrong with this lap joint. But if it's loaded beyond normal limits, failure is likely to occur in the heat-affected zone where grains are coarse and relatively weak (arrow). The torn surface will show a characteristic crystalline structure. There's not much that can be done, short of post-welding heat treatment to normalise the grain.

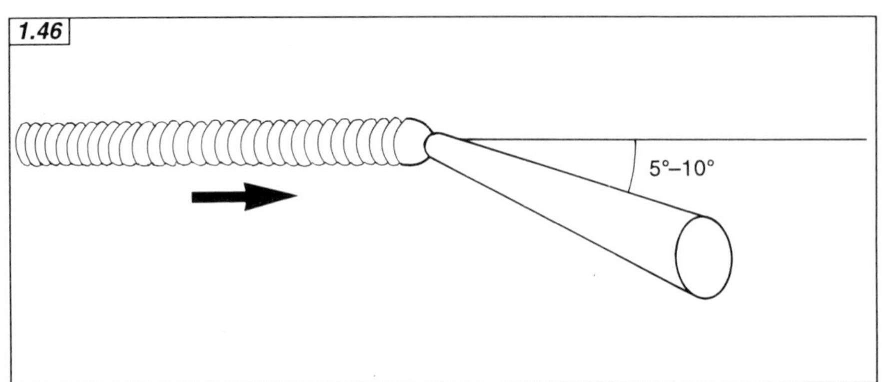

1.46 Welding along the side of a vertical plate can produce globs of metal to rival Babylon's Hanging Gardens. Using too much heat (current) or too-slow travel is usually the cause. Dial in only enough current to give penetration and a smooth-running arc. Changes in arc length lead to a 'humpy' appearance, as a too-long arc lets metal sag. Angle the rod as shown above to keep the bead pushed up into the joint, keep a constant shortish arc and don't travel too slowly.

1.47 Two jobs using a dissimilar steels electrode: a sash cramp's cast iron endplate welded to the central mild steel beam for more rigidity (A), and a slurry pump's cast steel shear plate resurfaced back to near-original dimensions (B, right).

silver. It's usually sold in short lengths.

Both black and bright mild steels are easily filed and give off long, light yellow sparks when touched by an angle grinder. Both are readily weldable, while silver steel is not.

Putting more carbon into steel makes it harder; Table 1.1 shows what happens. Carbon steels are the result, and are sub-divided according to their carbon content. As this grows, so do hardness, brittleness and the difficulty of welding. Such material is often heat-treated after forming to shape to boost its resilience.

This must be considered when welding. Springs are a classic case: the arc's heat will probably 'draw the temper', making the weld area hard, brittle and unsuitable for service until retempered.

As carbon content goes up, so steel gets harder to file. Indeed, files themselves are made from the highest carbon steels. Thus an unknown material can be checked against a mild steel standard by filing. If it won't file, it's probably not weldable.

Grinding spark pattern also changes. As the carbon content goes up, sparks are shorter, bush out closer to the grinding wheel and may be darker yellow in colour. If in doubt, compare an unknown with a mild steel standard.

Although heat treatment improves a carbon steel's toughness, still more gains come from adding small quantities of exotic elements. Thus are born the alloy steels.

Small quantities of all sorts of metals—nickel, tungsten, manganese, molybdenum, cobalt, vanadium—are cocktailed together with carbon steel to produce materials for a wide range of applications. The end result is usually heat-treated to maximise its properties.

Alloy steels turn up all over the place. Springs and transmission components, both of which usefully exploit the material's toughness and resilience, are common applications.

Stainless steel is another example, using a good spicing of chromium to resist corrosion. To the metalworker this is both good and bad news: for though it's slow to tarnish, stainless steel's very reluctance to oxidise means it can't be gas-cut.

Sorting an alloy from a carbon steel is largely a matter of application, though stainless stands out readily enough. Cost comes into it, too: a cheap hand tool is more likely to get its hardness from a tempered carbon steel than an expensive alloy mix.

Castings can be recognised by their complex shapes, generally rough surface finish and any raised surface lettering. But not all castings are iron.

Unless it has been heat-treated to a malleable form, cast iron's brittle nature restricts its use to areas where shock loads are low and compressive loads high. Bearing housings and belt pulleys are typical.

For harder service, cast steels come into their own. These are much tougher and can be heat treated for resilience. Thus they are used where a durable complex shape is called for.

Sorting out the two types of cast is a matter for the grinder. Cast irons produce few sparks, and these are an unmistakable dull red/orange red.

Cast steel will spark yellow—much like its mild counterpart—though sparks will be closer to the wheel and more bushy.

The hammer test usually works, too. Tap cast steel and it rings, while cast iron strikes a dull note.

Should you weld it?

Making 100% repairs in other than mild steel is a specialist business. The array of electrodes is bewildering, their selection important, and welding technique central to success. From this lot, one clear message emerges: *where any safety-related component is*

TABLE 1.1 Steel's hardness grows with increasing carbon content. But brittleness also increases, leading to the need for heat treatment to get round this. Steels in the lower reaches of the series are weldable on the farm, given appropriate rods. Ditto for those in the middle, though greater care in rod selection, joint and rod preparation and subsequent cooling is called for. High carbon steels are unweldable by normal methods. Adding dashes of other elements gives a range of tougher alloy steels—see text. Welding cast irons is beyond our brief, but if you're tempted, first identify the material and then ask advice on the rod and technique to use. Electrode companies are very helpful. Wherever safety is involved, *give the job to a specialist*.

Material	*Percentage carbon*	*Typical use*
Wrought iron	0.01–0.03	Ornamental ironwork: little used now
Dead mild steel	0.05–0.1	Wire, pressings
Mild steel	0.1–0.3	General engineering as strip, tube, plate
Medium carbon steel	0.3–0.5	Harder structural steels, railway lines, cast steels Weldable, but may be made brittle by welding heat
High carbon steel	0.5–0.7	Harder, also more difficult to weld. Will be made hard and brittle by weld heat. Chisels, springs, some hand tools
	0.9–1.5	Axes, hammers, files, razor blades. Very hard, effectively unweldable on farm
Grey cast iron	2.0–4.0	Castings not subject to shock; machinable
White cast iron	2.0–4.0	Very hard, brittle, not easily machined
Malleable cast iron	2.0–4.0	Comes from heat-treating white cast. Much more elastic. Some cutter bar fingers, coulter brackets, vice bodies, etc

Note: Blast furnaces make pig iron, which is high in carbon and impurities. Refining produces the steel series above; adding small quantities of other elements produces a range of tough alloy steels

involved, don't try to weld it. Take it to a Man Who Knows.

That's only good sense. But in non-critical applications, the day can often be saved by a 'dissimilar steels' electrode.

Although metallurgists will always stress the importance of matching rod and material, it's surprising what can be achieved with one of these jack-of-all-trades rods.

A few tips may help with difficult-to-weld material. For a start, choose a rod which will deal with the most awkward of the metals to be joined.

The work will probably need preheating, then cooling slowly to minimise the possibility of post-welding cracks. A muffle furnace is ideal, but who has one? A gas flame will usually do if heating is even.

Keep heat input low during welding. This means using no more current than is necessary for fusion, and minimising the number of runs. After welding, let the work cool very slowly. Keep it away from draughts and cold surfaces, and never, ever, quench-cool. Even with mild steel, avoid quench-cooling where strength is critical.

Arc welding cast iron is beyond our brief here, but a few words are in order. Preheating helps a great deal, and low welding heat input/ slow post-weld cooling are always necessary.

Success with cast iron is never completely certain thanks to the material's tendency to crack as it cools. Gas welding, either using a high-manganese bronze rod or fusion welding with a high-silicon cast rod, is an alternative to arc work. The former is a non-fusion process, so helps keep temperatures down.

In either case it's important to know which cast iron you're dealing with. Malleable cast will cool to brittleness if arc welded, though lower-temperature bronze welding can work. Similarly, grey cast will turn to the brittle white form if cooled too quickly.

Section 2
MIG/MAG WELDING

Now for the new boy. Once the sole province of industry (where it was developed to speed production work), equipment for MIG/MAG work has slowly drifted down in price until it's within the reach of many farms. Here we'll deal with sets most likely to be bought new by medium-sized farm workshops—single phase units using 0.6mm, 0.8mm and 1mm wires. Bigger sets shoot up in price and offer more output and control sophistication, though their basic operating principles stay the same.

So what's involved? In short, an arc welding process in which MMA's individual consumable electrodes are replaced by continuously fed wire, and electrode flux is replaced by an inert gas shield.

Terminology next. MIG is shorthand for Metal Inert Gas. The only truly inert (i.e. non-chemically reactive) shield gases are argon and helium, which are reserved exclusively for welding aluminium or other non-ferrous metals and their alloys. MIG welding must therefore refer solely to such work.

Here we're talking about welding steels, where either carbon dioxide (CO_2) or a CO_2/argon mix is the usual shield gas (pure argon makes for an unstable arc). As carbon dioxide is not inert, the process can't rightfully be called MIG. Metal Active Gas (MAG) welding is a more appropriate term, and will be used from here on.

Convert or beginner?

MAG welding has the reputation of being easy, and it is—if approached the right way. Most agricultural users will already be more or less proficient with MMA but it's best for the experienced MMA user to forget practically everything and start afresh. Applying MMA rules to MAG work doesn't really work.

The new technique stands most things on its head. For example, it's always a DC (direct current) process. The operator is no longer in manual charge of arc length: this is changed by altering voltage at the set. Current is adjusted indirectly by altering wire feed speed. Travel is faster and usually in the opposite direction from MMA, and there's effectively no help from flux.

The only really constant thing is the importance to weld strength of those Terrible Twins—penetration and fusion. These are unchanging, no matter whether a weld is by MMA, MAG, gas flame or mind over matter.

The benefits

MAG welding offers a fair selection of advantages over MMA. For instance:

- MAG's relatively small filler wires carry similar current to MMA electrodes only on thinner-section wire, e.g. 130A on 1mm-thick wire versus the same for a 3.25mm electrode. So there's great potential for excellent fusion and penetration as long as technique is good.
- MAG is faster, in both speed of travel and post-welding clean-up.
- MAG operation requires less manual skill, as arc length is controlled by the set rather than by the operator. Against this must be stacked the need for more careful setting of welding conditions—the 'tuning' of voltage and current.
- The weld area is easier to see, so potential problems can be sooner spotted.
- There's no heavy slag to control or to chip off after a run.
- Given a reasonable volume of work, MAG welding is potentially cheaper—even considering the cost of cylinder rental and shield gas.

The last point has to be looked at carefully in a farm context. MAG plant is considerably more expensive than equivalent MMA gear, thanks to its wire-feed mechanism and AC rectifier. Extra complexity means more potential for going wrong. Thus buying an expensive MAG set for occasional use makes little sense.

There are also jobs where MMA's flexibility over electrode type and (relative) contamination tolerance make it the better choice. Relying solely on MAG for the whole spectrum of farm work isn't a good idea. MAG with MMA backup is.

Given a reasonable volume of mild steel repair/fabrication work over a period of time, MAG's flexibility will pay off. It's faster, welds a wide(r) range of thicknesses and demands less skill to make a reasonable job.

MMA has been honed over the years into a flexible and reliable friend. But it has two not-insignificant drawbacks: it's relatively slow (rods have to be changed and welds de-slagged) and a high level of operator control is needed, particularly when joining thin sections.

MAG welding is free of these shortcomings, but at the same time introduces some of its own.

What's involved—more detail

The slowness of MMA is solved by replacing its individual rods by bulk electrode wire on a spool. The user holds a gun-like 'torch', whose trigger starts and stops wire and gas

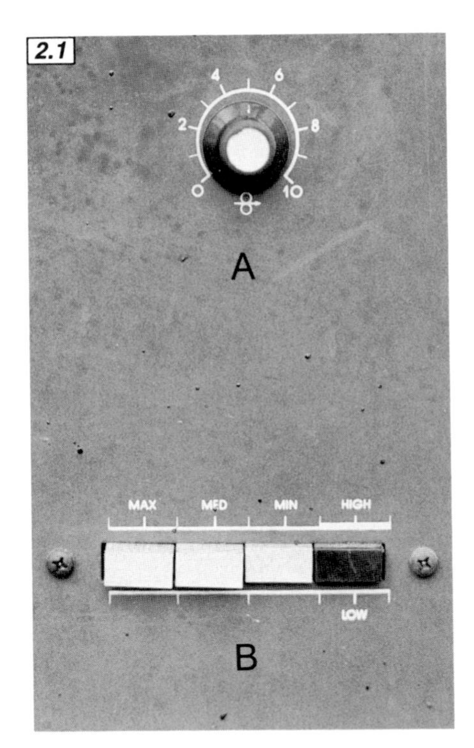

2.1 A single-phase MAG welding set's front panel. Its MMA brother has control(s) to change welding current, while the MAG operator can alter two things—wire feed speed (A) and voltage (B). Changing voltage alters heat input in big steps; changing wire feed speed fine-tunes it by varying current—see page 26.

feed. On all but the cheapest sets, this trigger simultaneously switches welding current on and off. Cheap sets have their welding wire 'live' all the time, so the risk of accidental arcing is ever-present.

Wire is power-fed continuously to the weld pool by a motor and roller arrangement. Wire speed is looked after separately by a control on the set (Fig 2.1).

Wire passes through a copper contact tip as it leaves the gun, picking up power en route. A copper coating on the wire both improves electrical contact and delays surface rusting.

That's filler material briefly dealt with. Replacing the cleaning and gas shield action of MMA electrode flux is harder, and requires a two-pronged attack.

Compressed gas is used to shield the weld pool. A regulator or flow meter on the supply cylinder looks after gas pressure and volume, while flexible tubing takes it to the torch (Fig 2.3). Here a nozzle (usually adjustable) directs gas flow over the weld area, keeping off atmospheric contaminants (Fig 2.4).

Duplicating the chemical cleaning action of MMA flux is harder. The filler wire includes small quantities of deoxidants—silicon, manganese and perhaps aluminium, titanium and zirconium. But these can't match a full flux coating for efficiency, which is why MAG welding is much more contamination-sensitive than MMA.

The front panel carries at least one more control than an MMA set. In addition to an on/off switch, there are either push-buttons or a stepped control for voltage, and a second knob for wire speed. More complex sets offer functions like spot-weld timing and pulse control.

The difficult bit

To get the most from a MAG set, it's necessary to dig deeper.

Some general points to start. A long electric arc takes more voltage to sustain it than a short one. Why? There's more electrical resistance in a long arc, which takes more volts (electrical pressure) to overcome. As arc length grows, so current drops, thanks to that increased resistance.

Power supply design is different for MMA and MAG sets. It's this that makes it possible to control the two forms of welding.

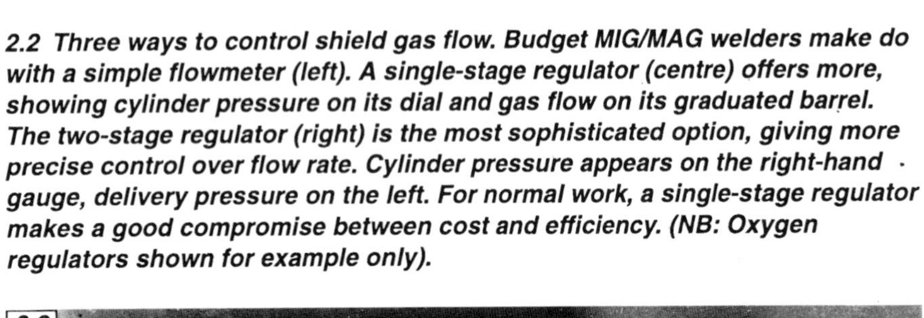

2.2 Three ways to control shield gas flow. Budget MIG/MAG welders make do with a simple flowmeter (left). A single-stage regulator (centre) offers more, showing cylinder pressure on its dial and gas flow on its graduated barrel. The two-stage regulator (right) is the most sophisticated option, giving more precise control over flow rate. Cylinder pressure appears on the right-hand gauge, delivery pressure on the left. For normal work, a single-stage regulator makes a good compromise between cost and efficiency. (NB: Oxygen regulators shown for example only).

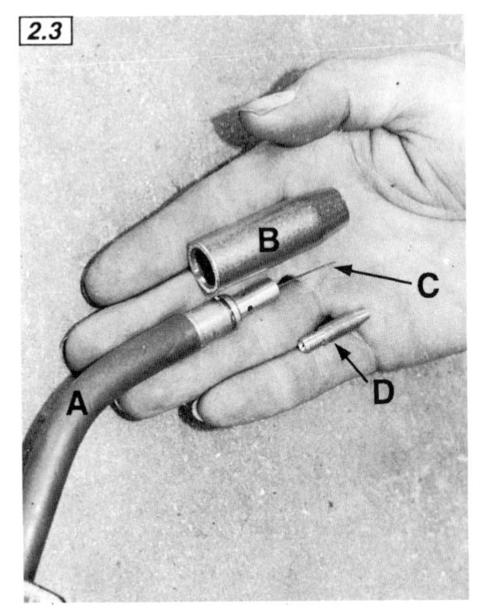

2.3 Shield gas flows from the gun (A) before passing through the nozzle (B), and then to the weld area. Filler wire (C) is fed to the weld pool mechanically via the torch liner, picking up electrical energy from a copper contact tip (D) as it leaves the gun.

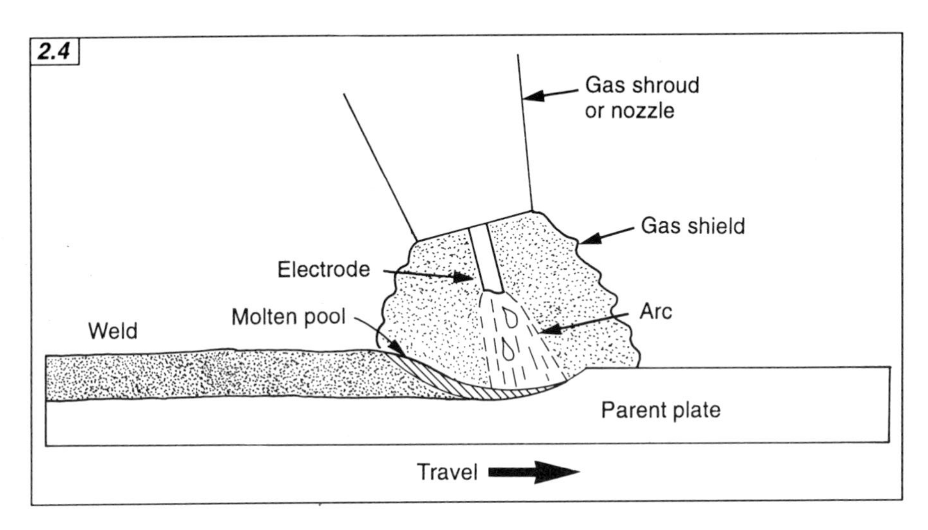

2.4 MAG in action. Shield gas from a cylinder replaces MMA's flux-generated protective envelope, keeping air away from the molten pool. Mechanically fed wire forms the electrode. Filler metal can be transferred to the molten pool in various ways, and this carries practical implications. For details see page 26.

largely dependent on the operator's ability to feed rod steadily to the joint and to spot and counter changes in arc length.

What's different with MAG? It uses much thinner filler wires which burn back much faster. At the rate this happens, maintaining a constant arc length would be near-impossible for a human operator.

MAG equipment overcomes man's limitations by delivering a *self-adjusting arc*, whose length is controlled automatically.

This small miracle is achieved by reversing operation of the power supply's 'elastic band'. MAG sets try to keep voltage constant as welding current changes. The operator chooses voltage level on the set's front panel, and the power supply does its best to stick to this; see Fig 2.6.

On the familiar AC-powered MMA plant, the user adjusts welding current to control heat input, then fine-tunes this manually by playing with arc length. The set's voltage output is predetermined, being fixed at one or more levels depending on the equipment. 50V and 80V lines are common—see page 2.

For consistent fusion and penetration, current needs to remain pretty constant during work. But the MMA welder's unsteady hand means that arc length is always varying, though maybe not by much.

As arc length increases, voltage goes up. As it shortens, voltage drops. Welding current wants to change too.

Power supply design can minimise this change. MMA sets are designed so that a large alteration in arc length (i.e. voltage) produces only a small change in current, minimising the effect of operator unsteadiness on welding heat. It's as though the set is acting as a piece of elastic between voltage and current—see Fig 2.5.

The relatively thick MMA electrode burns back fairly slowly, giving the operator reasonable time to react to (and counter) changes in arc length. But if he fails to do so, the power supply does what it can to help by holding current as steady as possible, minimising the effect on weld quality. Despite this help, delivering a good MMA weld is still

2.5 Power supply characteristic for MMA. On striking the arc, voltage falls from the open circuit value (A) to the running voltage band (B). Alterations in arc length produce voltage variations in this band; see how the curve's shape produces only corresponding small changes in welding current (C). Thus the set is actively helping to maintain current—and hence weld quality—when the operator's arm wobbles. Small current changes mean small changes in burn-back rate of a relatively thick electrode, giving the user time to spot (and counter) shifts in arc length.

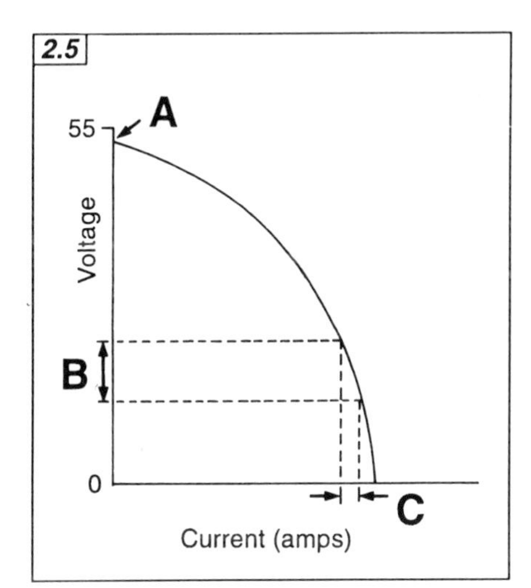

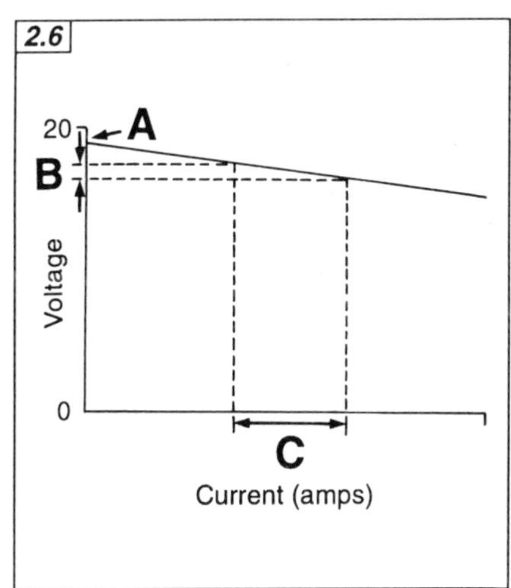

2.6 In contrast to Fig 2.5, a MAG power supply holds voltage relatively constant (B) for a large change in current (C). So arc length is automatically held steady as it tries to shift, making a MAG set much easier for the inexperienced to use. Why is this necessary? MAG wire is much thinner than an MMA electrode, so burns back much quicker at a given current level. As even Superman couldn't react fast enough to compensate for the resulting shifts in arc length, automatic control is vital. On a different tack, MAG's skinny wire carries very high current density, so an arc establishes readily at a much lower OCV—check out value (A).

The result is an arc that stays largely constant in length, no matter what the operator's arm does. Raising the voltage gives a longer arc, which fans out over more of the weld area.

By itself, this longer arc won't do a lot of good. All that happens is that the filler wire is burnt closer to the contact tip, and will eventually fuse with it. As a countermeasure, wire speed needs to increase with voltage. This also raises current (see below), giving more heat for fusion/penetration. Most sets link the two automatically.

MAG's thinner filler wires carry high current density, letting an arc establish with fewer volts. So single-phase MAG sets run at 17–25V, with intermediate values selected stepwise.

There's a safety spin-off here. As MAG voltages are well below MMA's 55 or 80V OCV output there's much less risk of a 'tingle' from a MAG set on damp days, and its DC output is less readily felt.

The MAG trick

Miss this bit if electrics confuse you. Just how does the power source produce a self-adjusting arc?

In a MAG set, mechanical wire feed is taking the place of the MMA operator's downward-moving arm.

Should a change in wire feed speed or operator positioning try to lengthen the arc, circuit voltage rises rapidly in response. The power supply reacts by dropping current substantially; see Fig 2.6.

At this lower current the filler wire no longer burns back so quickly, but it's still being piled into the weld area at undiminished speed. So the arc very quickly shortens, taking length back to where it was before the disturbance. Self-adjustment happens so fast that the operator doesn't notice any change.

The reverse applies when a change in feed speed or operator position tries to shorten the original arc. A small voltage drop produces a big rise in current; the wire burns back faster and the arc returns to its original length. It's as though the arc is held suspended between two rubber bands: pull one, and the other acts to take length back to where it was.

Current change

Changing voltage at the set's control panel alters arc length. But how is current varied? By changing wire feed speed. Reducing wire speed at a pre-set voltage means that the wire is burnt back further in any given time. Average arc length goes up a little, voltage increases, and current drops. So lowering wire speed reduces current.

The reverse happens when wire speed is increased. Average arc length shortens a bit, voltage drops and current rises.

Thus with MAG sets, *current is adjusted by altering wire feed speed —up for more current, down for less.*

When voltage is altered, most sets automatically rescale wire speed range to suit. This ensures that current roughly matches the new voltage, but it's up to the operator to bring the two into proper balance by fine-tuning speed. See page 31 for procedure.

But the operator has only limited scope for current adjustment. If he reduces speed too much, burn-back takes the arc back to the contact tip. Wire and tip will probably fuse and stop the feed.

On the other hand, if he increases speed more than burn-back can cope with, unfused wire jabs repeatedly into the weld pool. He'll feel this as a judder or 'stubbing', and weld quality will go down.

So to sum up. Voltage has by far the most effect on arc length, whereas wire speed alteration affects it only a little. At any given voltage there is a range of current available, but it isn't great.

A Quick Recap

Practical points to emerge from the above theoretical excursion are these:

- With MAG, the operator pre-selects voltage and hence arc length.
- The more voltage, the longer the arc. Once wire feed speed is increased to suit, total energy input is greater. Thus higher voltages/speeds are used for welding thicker sections. Lower voltages mean shorter arcs and less energy, so are used on thinner sections.
- More energy input increases penetration and fusion in a given material thickness. Within limits, the operator can change energy input by increasing or decreasing current.
- Current is changed by altering wire feed speed. At a given voltage setting, the faster the feed, the higher the current—and vice versa.

All this suggests that the operator has more options over settings than with MMA. What are those choices?

Which process?

First, the operator may be able to choose how metal ends up in the weld pool. The options actually depend on the set's output and the wire size/shield gas combination used. Where options exist, they carry great implications for penetration and fusion, so we'll look at each in turn.

Most normal farm jobs—joining sheet, thicker sections and filling holes, both on the flat and positionally—call for a well-behaved, relatively cool and quick-freezing weld pool. A MAG set delivers this when operating in **dip transfer** mode.

Dip transfer happens at relatively low voltage/current settings. Wire arrives at the weld area slightly faster than it burns off, so its tip actually dips into the molten pool. There's a short-circuit and the arc goes out immediately (Fig 2.7).

Current rises sharply, heating the wire enough to melt it. The short-circuit is broken and the arc restarts, leaving a globule of molten metal behind in the weld pool.

Dip transfer happens with both CO_2 and mixed shield gases, and is easily recognisable. The very rapid short-circuit/re-establishment of the arc produces an unmistakable crackling noise; listen to a properly adjusted set to hear its ripping-cloth sound.

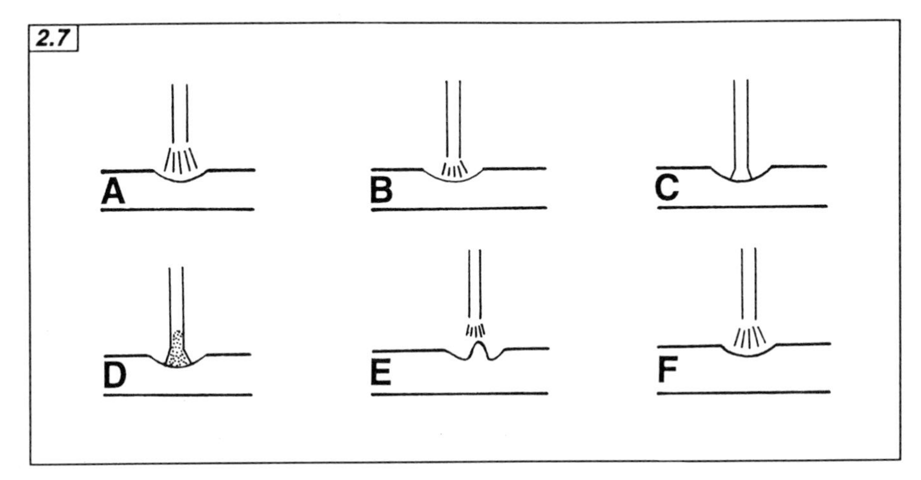

2.7 What's happening during dip transfer. (A) An arc exists between wire tip and weld pool. (B) Wire is fed faster than the arc is burning it back, so wire gets closer to the pool. (C) Wire tip and pool eventually meet, and the arc goes out. Dip is also called short-circuit transfer. (D) The resulting short-circuit resistance heats the wire, and this heat travels upwards. (E) Very soon the wire gets hot enough to melt, leaving behind a globule in the weld pool. The arc then reestablishes, taking the cycle back to square one. (F) This dip–extinguish–reestablish pattern repeats itself many times per second, giving some spatter and MAG's characteristic crackle. (NB: For clarity the contact tip, nozzle and shield gas are not shown.)

Dip transfer is the norm for low-output sets, and for bigger units working positionally or with thin material. As no molten metal crosses the arc, vertical and horizontal/vertical welding is relatively easy.

But it's not the only option. Where the job allows and the plant can muster sufficient current for the wire/shield gas used, the operator can opt for **spray transfer**—Fig 2.8.

In this process, energy output in a given time is higher. Why? Because the arc never goes out. Penetration is deeper and more weld metal is deposited. Thus it's a useful option for fast, strong work in relatively thick-section steel, but is too 'fierce' for thin materials.

On steels, spray transfer is only usable on the flat and in the horizontal/vertical position. The weld pool is too hot and too liquid for other work. (NB: Aluminium is all-position welded by spray transfer, but that's a different matter.)

Moving into spray mode calls for more current, and the wire burn-off rate is much higher than in dip transfer. After arc initiation the wire's tip never has a chance to short-circuit in the weld pool, so the arc never goes out. Total energy release is therefore greater.

2.8 In spray transfer there's enough energy around to balance wire feed, so the tip and pool never meet. Small molten globules actually fly across a long arc, which never goes out. Energy release per unit time is greater (giving better penetration) and spatter is less.

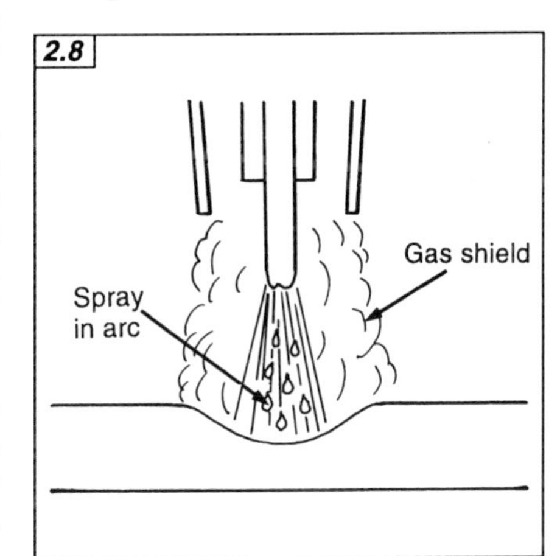

Molten metal crosses the arc continually as a spray of tiny droplets—hence the process name. Filler wire is burnt off very fast, so the set must be able to deliver fresh supplies at a rate of knots.

There are other differences from dip transfer. Tiny droplets move to the weld pool in an orderly fashion (thanks to strong magnetic fields around the high-current arc) so there's little spatter.

The arc both sounds and looks different, too. There's no longer any 'crackle' to it, as there's no short-circuiting. In spray mode, arc length grows (thanks to the higher voltage used) and its sound changes to a steady, quiet hum; light output goes up and the weld pool widens. You can't miss it.

There's no clean-cut changeover between the two modes, as the voltage and current at which transition occurs depend on both the shield gas and wire thickness. As spray transfer moves back towards dip mode with falling arc energy, **globular transfer** (a cross between dip and spray) occurs. Much larger droplets transfer across the arc, causing considerable spatter as they hit the weld pool.

Small sets can't muster enough current to allow spray transfer to start. But single-phase sets delivering 180A or more should do it, using currents over 160-170A on 0.8mm wire and argon/CO_2 shield gas. Thus this mode of operation is a possibility for mid-size farm equipment. If it's used, keep an eye on the torch—spray's extra heat release can cook it.

Whether spray transfer is a viable option depends on two other factors. Wire diameter is one: fine 0.6mm filler needs to be fed really fast to satisfy the arc, and the set may or may not be able to deliver the quantity required. Very high feed speeds increase the chance of a snarl-up. Progressively thicker wires need less rapid feed (but increasingly more current) to weld in spray mode.

Shield gas also plays a big part. True spray conditions can't be achieved with pure CO_2—only with an argon mix. See page 32 for the way to dial in spray transfer.

Which wire?

Individual makers have their own trade names for MAG filler wires—e.g. the Bostrand family from Murex and Oerlikon's Carbofil range.

No matter which name it goes

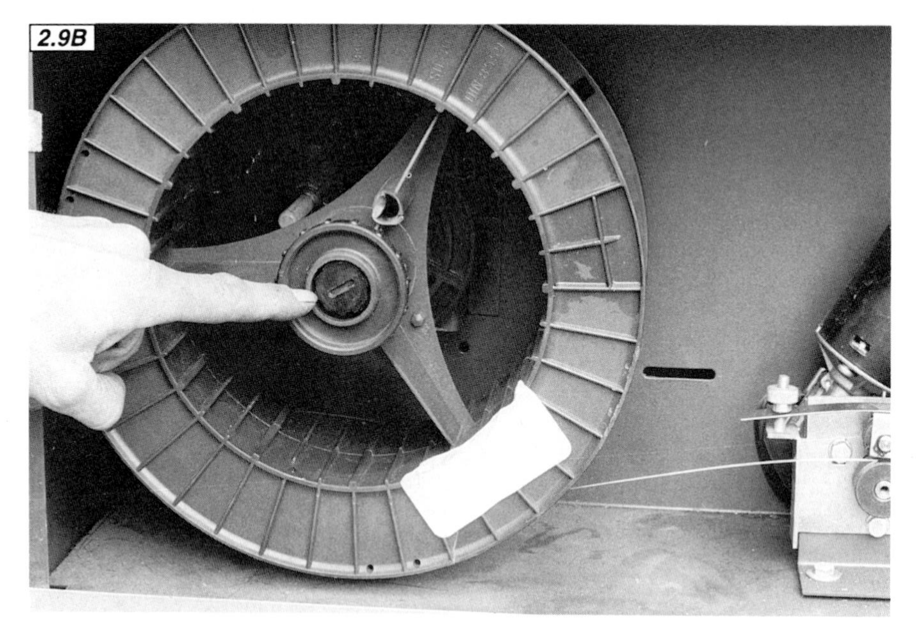

2.9 After several days of winter damp, unused wire is likely to pick up a thin, often patchy layer of corrosion—especially on the top coils. Using steel wool and a piece of rag in series helps produce a smooth feed and consistent work (A). Don't clamp the wire too tight, and change the materials occasionally. Check spool friction drag (B) and feed roller adjustment (C) if wire feed is erratic (see page 30).

under, for minimal cost fabrication/repair in mild and low-alloy steels, double deoxidised wire to grade A18 of the appropriate British Standard is fine. And unless you specify otherwise, that's what your supplier will probably provide. If you don't mind paying a little extra, precision-wound spools give a smoother feed than the normally supplied random-wound versions; the price premium was around 17% at the time of writing.

An alternative is triple-deoxidised wire to BS grade A15. This handles contamination better, but at around 2½ times the price of A18 it's significantly more expensive. Other wire grades are available for high-alloy and stainless steels, but don't concern us here.

Most single-phase sets handle at least two wire diameters (typically 0.6mm and 0.8mm), and it pays to keep stocks of both. The thicker wire will be usable on all general work (including sheet metal) but a change to 0.6mm is usually needed for very thin sections.

Wires are generally copper-coated to resist corrosion. But prolonged damp will affect performance—wire with even light surface rust won't feed evenly and produces lower-strength welds. If the set won't be used for several days it's best to take off the wire spool and place it somewhere dry. Spare spools should always be kept that way.

Clean filler wire always feeds and welds best. An arrangement with wire wool, cloth and a couple of clips is simple insurance (Fig 2.9A). This cleans wire as it passes, helping feed rollers get a grip and minimising contaminant build-up in the guide tube. It's surprising how much crud is picked off.

It's a gas

For mild steel, either CO_2 or a CO_2/argon mix can be used. A run through their strengths and weaknesses might help with choice.

CO_2 gives narrower, deeper penetration and a less fluid molten pool that's easier to control on positional work. It's also cheaper and longer-lasting per cylinder.

On the downside, a heater may be needed to overcome regulator frosting, particularly in winter. CO_2 is more demanding as far as plant setting goes and it produces a 'rougher' finished weld with more spatter. The arc is less smooth-running, and substantially higher currents are needed to move away from dip transfer than with a gas mix. Some globular transfer always occurs, which means that spatter can never be 'tuned out' by the operator. Ideally, triple-deoxidised wire should be used with CO_2 shield gas to maximise weld strength.

By contrast, an argon/CO_2 mix promotes a hotter, more fluid molten pool and supports true spray transfer. Penetration is broader but shallower, making it better for thin-section material. Spatter is reduced, the weld surface is smoother, the arc more stable and plant setting is less critical.

Gas mixtures containing 5% or 20% CO_2 are available, though not necessarily in all cylinder sizes. With either dip or spray transfer, the 20% mix gives a slightly cooler pool better suited to positional work.

So which to go for? For general farm work, a 20% gas mix such as BOC's Argoshield 20 is good. Though more expensive than CO_2, it's easier to work with over a range of thicknesses. A 5% mix is best suited to thin sections. Cheap sets are normally designed with a mixed shield gas in mind.

Getting started

The filler wire and shield gas have both been chosen. Before getting underway, run through preflight checks:

1. Check contact tip condition (Fig 2.10). Welding current is picked up here and good, consistent wire/tip contact is vital for a smooth arc. If the wire flops about in a used tip, replace it.

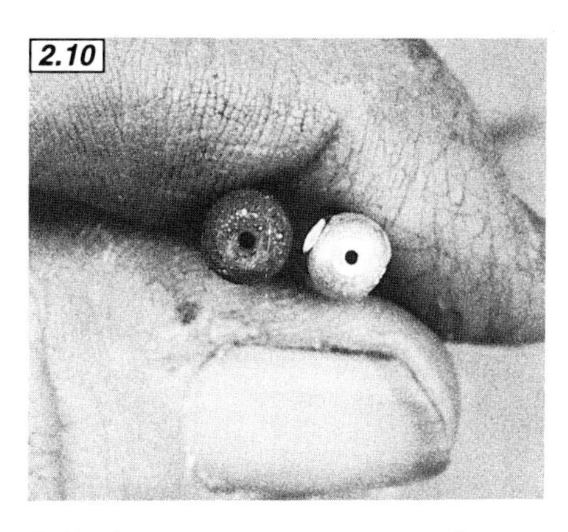

2.10 Compare a worn contact tip (left) with a new one. Tips wear out in time, so inspect frequently and change as soon as wire becomes a floppy fit. If you don't, current pick-up will soon become intermittent and the arc won't run steadily.

If in doubt, try a length of the next size up in filler wire. If this goes through a used contact tip, throw that tip away. Be very careful if trying to use gas nozzle reamers to reclaim a tip that's been part-blocked by spatter; the soft copper will usually be reamed out before the contaminant.

2. Take a look at the nozzle (Fig 2.11). The more this crusts with spatter, the less shield gas can flow. Serious crud can bridge the gap between nozzle and tip, making the nozzle live and possibly bringing on arcing between it and the work/surroundings. Shift deposits with a carefully wielded blade; spatter-release liquid (sprayed on before welding starts) helps a lot. Looked-after nozzles last a long time—throw one out only when it's eroded away or badly knocked about. During work, check nozzle condition frequently.

2.11 Several minutes' dip transfer work at high current gives a crusting of spatter round the nozzle, particularly when working vertically or overhead. If left, this will eventually cut gas flow and perhaps bridge to the contact tip, making the nozzle 'live'. Inspect and clean frequently.

3. Where possible and appropriate, adjust nozzle height to suit the metal transfer process (Fig 2.12 A and B). For normal dip transfer

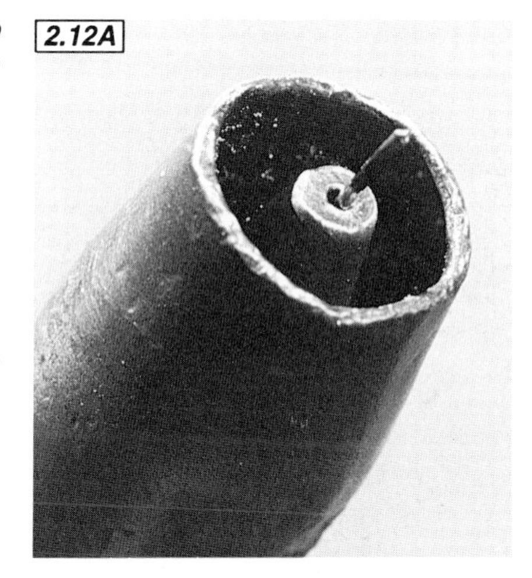

2.12 For dip transfer, set the nozzle so it's level with or just projects beyond the contact tip (A). This minimises electrical resistance (see Fig 2.19) and gives the best view of proceedings. Spatter build-up will be fast, though. For spray transfer, set the nozzle so it masks the tip by about 6mm (B).

work, the contact tip can be level with or just projecting beyond the nozzle's end. This minimises electrical resistance caused by long stick-out (Fig 2.19), and simultaneously maximises both welding current and the operator's view of the proceedings. In some cases working into a corner can be made easier by exposing the contact tip, but spatter build-up will be faster. The same applies when working vertically or overhead, so a good watch needs to be kept on tip condition. Be careful in dip work: the more the tip is recessed, the more is the tendency to make up for lost current by increasing wire speed. More wire than necessary is then used and penetration suffers.

For spray transfer work, draw the nozzle forward so it shrouds the tip by about 6mm—Fig 2.12B. This helps keep the tip away from heat and provides better gas protection for spray's wider arc and weld pool.

4. Spare a thought for the torch liner. Though not required often, an occasional blast-out with dry compressed air clears its tubes and makes for smooth wire supply. Use no oil!

If cleaning is neglected, wire can actually jam solid. Cheap sets may come with a Teflon or plastic liner (allowing use with aluminium filler wire), but a spiral steel item is a much longer-lasting bet for steel wire. Properly looked after, a good liner lasts a long time.

5. The spool-holding spindle may have an adjustable friction control (Fig. 2.9B). Set this to keep a light drag on the spool so it won't over-run when the feed motor stops. Too much friction strains the system, and can mean a jerky feed as rollers slip and slide. If the set is to be transported, increase friction or bumping can produce a 'bird nest' of wire round the spool.

6. Check feed roller pressure (Fig 2.9C). Too much can damage the wire, while too little means an uneven feed. The latter shows up as an unsteady arc and (perhaps) in the tendency for wire to burn back and fuse with the contact tip. Back off roller pressure until the drive just starts to slip as you brake wire between finger and thumb; then increase it until slipping stops. Keep feed rollers clean and free from oil.

7. Set shield gas flow. Rates are specified (typically 10 litres/min, or 3–5psi), but in practice it's enough to use your ear alongside the nozzle. With wire speed set at zero (lest you get an earful of wire, and also to avoid waste), operate the gun's trigger and increase flow at the regulator until a steady, soft hiss is heard. Then, once work has started, decrease flow as much as possible. Too little shield gas gives the weld a rough, mousse-like look (Figs 2.13 and 2.38). Contamination in the weld zone can vapourise and pock-mark the weld in a similar way.

To counteract draughts increase gas flow a little. But in normal work, using a little more gas 'to be sure' only wastes it. Shield gas has no effect on the arc's ability to handle surface contaminants, so there's no point in using more if the arc isn't behaving. Stop, find out why and rectify. You run the risk of lowering protection via high flow rates (especially with CO_2). The stream becomes turbulent, drawing in the very air that you're trying to exclude.

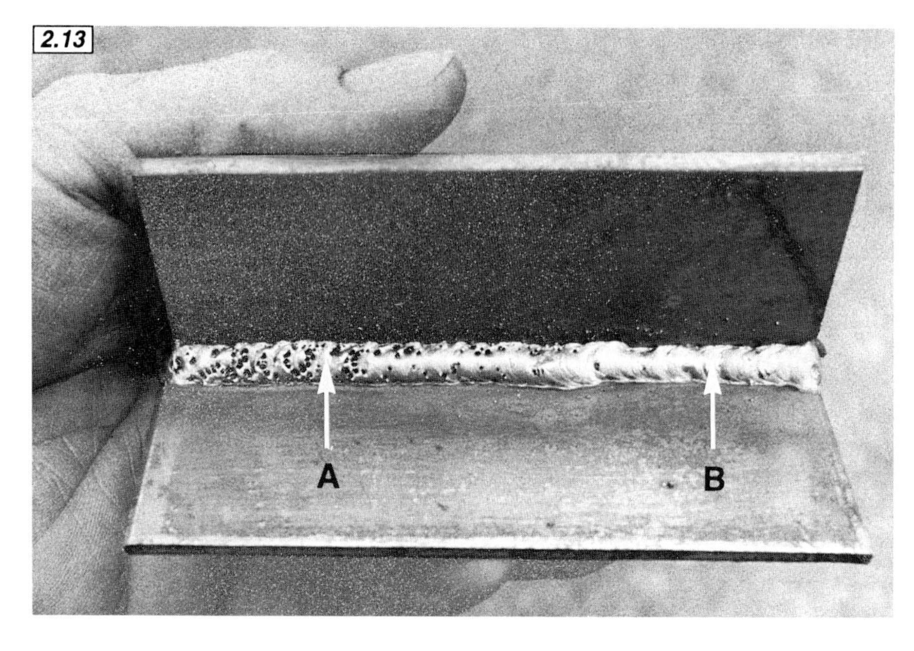

2.13 A lack of shield gas leaves a weld that looks like a rough 'Aero' chocolate bar and has about the same strength (A). This CO_2-shielded weld improved as flow was gradually restored to normal (B); contamination can give the same effect. When gas is really lacking (if you've forgotten to switch it on!) the arc tells you—it's very harsh and the weld pool boils. (See also Fig 2.38, page 43.)

Setting the Set

How do you choose the best combination of voltage and wire speed for the job in hand?

By experiment and experience—and the latter is quickly picked up. In the first instance we'll assume little or no familiarity with the welding set used, and consider dip transfer work.

There is no single 'best point' working condition to be found. Rather, there's a narrow working range which produces the soundest weld (i.e. one with adequate penetration/fusion and no defects), while minimising wire and shield gas use.

The following procedure is designed to let a new operator home in on this range. It may take some time, but if carried out properly with equipment in good repair it'll always bring results.

The basic principles are these:

1. Choose a voltage.
2. Optimise wire feed speed for that voltage and assess fusion. If it's acceptable, go ahead and weld.

3. If not, repeat the procedure at a different voltage.

During trials, stick to good scientific principles and alter only one variable at a time! Changing both voltage and wire speed at once produces only confusion.

Trial and not much error

Begin by selecting a voltage around mid-point on the set's scale. For thin material choose a setting a step or two below this; for thicker material start at mid-way or just above. When in doubt, go for mid-scale.

Next set wire feed speed to the mid-way point.

Find a piece of scrap metal equivalent in thickness to the intended job, and make a test run of at least 75mm on it. Travel a little faster than you would with MMA, but not so fast that the weld bead is 'stringy'—see page 35.

Look at the shape of the weld bead in the second half of the run. Its shape indicates the extent of fusion.

The chances are that the bead is a little humped and not fused well at its edges (Fig 2.14). Fine-tuning may sort this out.

Reduce current by lessening wire speed a little and make another run, keeping travel speed the same as before. The bead should look flatter and have better edge fusion: what you're doing is climbing the curve of Fig 2.15.

Repeat the process with progressively less wire speed each time, listening to the arc and checking bead shape after each run. The bead will tend to get flatter as less filler wire is added, fusion will improve as arc energy balances wire supply and spatter will increase as dip transfer

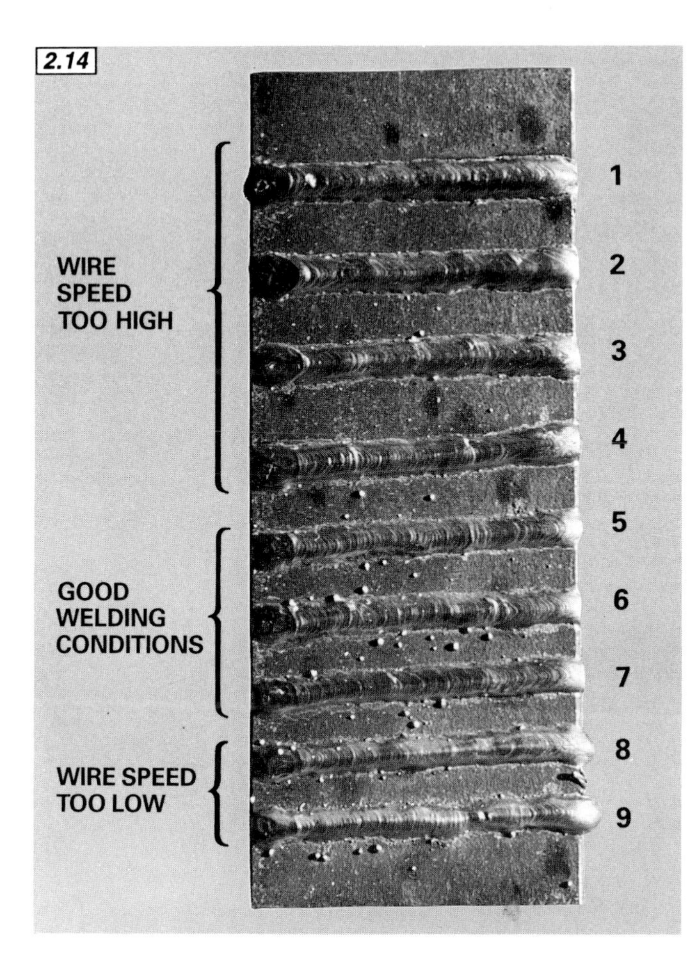

2.14 A series of runs at the same voltage and as near as possible at the same speed of travel. Wire speed was reduced between each. Starting at the top, runs 1–4 show heavy bead deposits characteristic of too-high feed speed. Runs 5–7 are in the 'optimum' range—the bead has flattened and spatter increased as speed dropped. The arc was crackling, but not as harshly as before. By runs 8 and 9 wire feed was too slow: fusion and bead regularity both suffered and the arc ran increasingly intermittently. Optimising wire speed for a given voltage produces maximum fusion and minimum wire use.

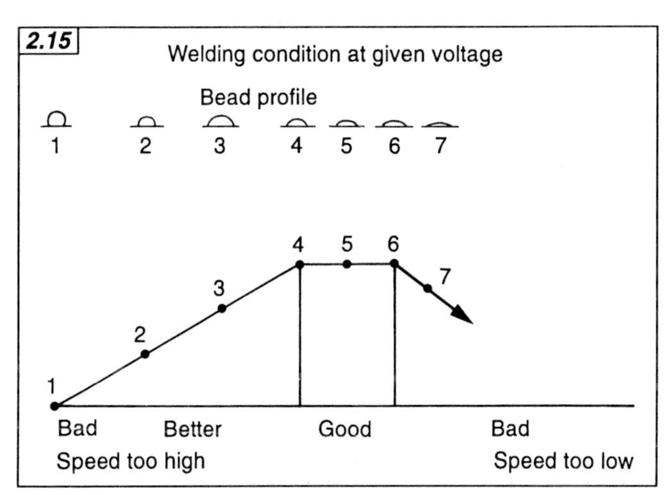

2.15 The theory behind Fig 2.14. At any given voltage, optimum welding conditions are found in a narrow band of wire feed speed. This example shows the idealised bead profiles of seven test runs, holding voltage and speed of travel constant but decreasing wire feed. When speed is too high for the available voltage the bead looks 'humped' on the plate, and its edges are poorly fused (1). Decreasing speed brings wire supply and voltage into better balance, so the weld pool looks more fluid, edge fusion improves and spatter goes up (2–4). The bead flattens only a little with decreasing speed in the 'optimum' range, fusion is good and the arc sounds less harsh and 'rippy' than before (4–6). But reducing speed too much (7) stops steady dip transfer operation. Metal is transferred as big blobs and the arc sounds spluttery. From this speed, an increase should restore good conditions. If the set's mechanical parameters are okay (feed mechanism, contact tip and gas flow) it should react to quite small changes in feed; MAG is very adjustment-sensitive. Aim to work in the range 4–6.
NB: The graph applies only where initial wire speed was too high. This is a very common condition, and is likely to apply when using the 'start from mid-range' approach to speed setting. If initial speed is too low, you'll find a spluttery arc with wire tending to burn back to the contact tip.

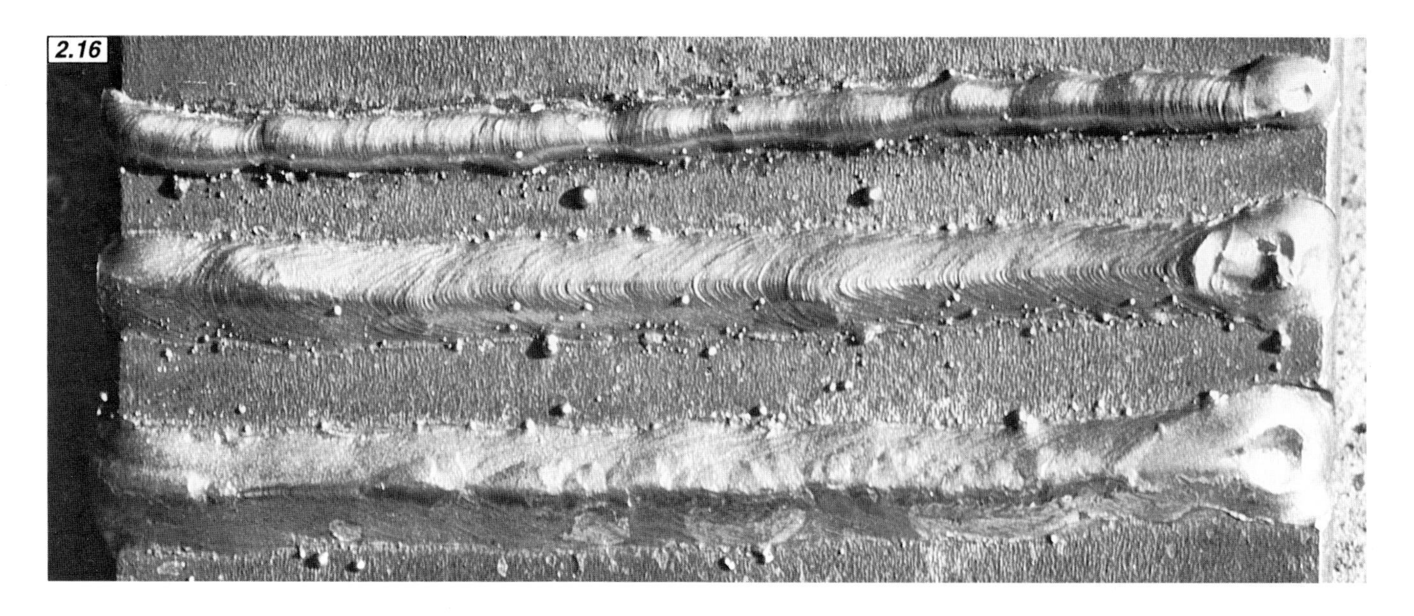

2.16 Changing voltage changes energy input, and hence fusion and penetration. When voltage is too low for the section to be welded, the bead is pinched, skinny and humped on the plate with little fusion (top). Increasing voltage gives enough heat for the bead to fuse into the plate (middle), and spatter goes up. But go too far and the bead widens and flattens, with fusion wanting to get out of control (bottom). The tendency then is to speed up travel, so the bead becomes ridge-backed with its ripples pulled out into Vee-shapes. Edge undercut appears, and burn-through will happen if the underlying material isn't thick enough to absorb the heat.

improves.

Arc noise will change too, softening slightly away from its original machine-gun crackle.

But quite suddenly, things will go downhill. Once feed is too slow the arc's length fluctuates, wire tends to burn back towards the contact tip and metal is transferred as blobs. Arc noise turns into a splutter, fusion worsens and the bead becomes uneven.

Once this happens, turn up wire speed a little. If the arc stabilises and runs smoothly, bull's-eye!—you're right in the 'good' band. Welding conditions are optimised for maximum fusion and minimum wire usage, and you're at the lower end of the usable range. If a little more current is needed, wire speed can be increased—but watch that the bead doesn't get humpy.

(Note: Significant changes in bead shape for small changes in wire speed may not appear. On some sets, turning the speed control through full scale only produces a relatively small change in feed. In such cases the effects above become apparent only for a large change in speed setting.)

If all the above works out successfully at the original voltage, then fine—your first guess was good.

But the original voltage choice may have been way out. Where burn-through is a problem which can't be cured by lowering wire speed, go down one voltage step and try again. Similarly, if good edge fusion can't be produced by increasing wire speed, go up one voltage step and re-optimise feed rate.

Where burn-through continues even at the lowest voltage setting, go down one wire size (where possible) and start again from the voltage mid-point. In extreme cases, stickout can be varied to control heat input (page 34).

The foregoing may seem the long way round, but it's the only way to start if you're uncertain which settings to use. After a while you'll go straight to a good approximation of voltage and find optimum wire speed with very little experiment.

Optimising for spray

So far we've been talking dip transfer. Whether it's possible to use spray transfer depends on your set, the shield gas used and the wire diameter—see page 27. You'll need at least 160A, 0.8mm or 1mm wire and argon/CO_2 shield gas.

Dialling-in spray is a little different than for dip. First things first: adjust the gas nozzle so it shrouds the contact tip by about 6mm, thus keeping transferred metal away from atmospheric contaminants. And, if such a thing is available, switch to a straight-sided nozzle—this produces a wider gas envelope to match spray mode's longer and wider arc.

To see if the set can produce spray, select maximum voltage output. Spray may be generated at lower settings—it all depends on the set's actual capability—but start at the top.

Optimise wire speed for the voltage selected, and check that the arc runs with dip transfer's characteristic 'crackle'.

Now reduce wire speed a little and try again. Listen and watch: repeat the process until the arc has just lost all its crackle. It'll be significantly longer than it was with dip transfer and will run with a soft, smooth hum, and spatter will die away. That's spray transfer working.

Check penetration and fusion. Both will be better than with dip transfer (Fig 2.17). If they're excessive for the job, try a lower voltage.

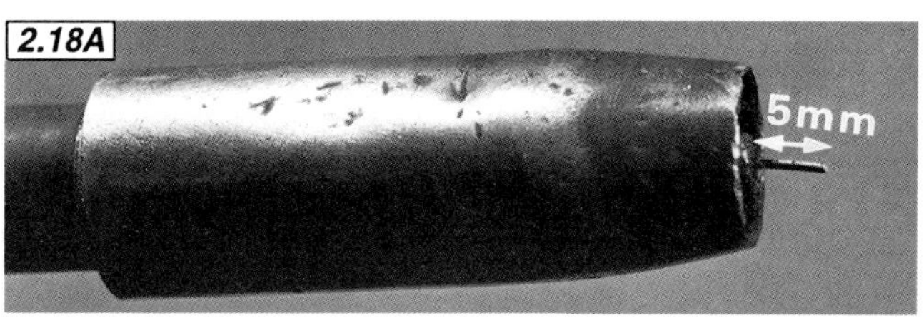

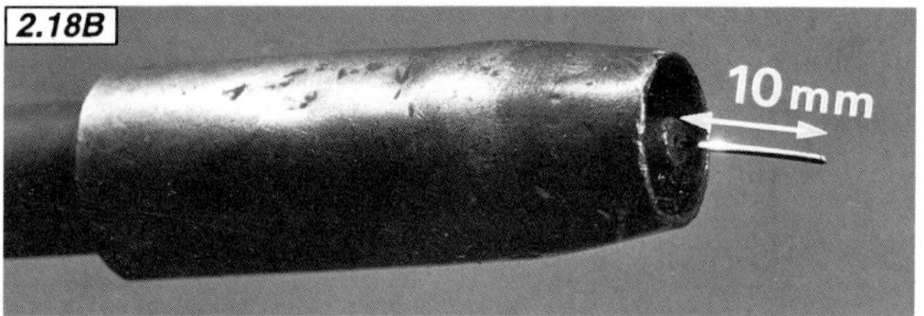

2.17 A small subsoiler wing (right), welded in the horizontal-vertical position using spray transfer and 0.8mm wire. A 180A set running 0.8mm wire was used, so spray was chosen to counteract the foot's large heat sink.

2.18 Normally use around 5mm stickout for dip transfer work (A), rising to a maximum of 10mm (B). Here the nozzle is set level with the contact tip. For a good view of thin-sheet work, it can protrude slightly if required.

But you can only go down so far and maintain true spray conditions. When voltage drops too low for the wire in use, transfer will move back to globular mode. Wire will melt in large dollops, spatter will increase dramatically and the arc will splutter.

Tidying up

In either dip or spray transfer the arc can wander uncontrollably between the wire and work. This is 'arc blow', and it's typical of DC processes—strongly directional magnetic fields are pulling the arc out of shape.

Arc blow is particularly noticeable when working in a tight corner, like welding the inside of a 90° joint in angle iron. Minimise the problem by keeping the welding return lead's clamp close to the weld area and working directly towards or away from it .

To finish the fine tuning process, look at shield gas flow. Reduce it a little and monitor the arc and weld quality; settle for just above the minimum needed in ambient conditions.

One last point. While it's okay to alter wire speed while the arc is running, don't change voltage. There's a good chance that you'll blow the set.

Various variables

That's adjusted the equipment. Now to do the same for the operator.

Both plant setting and technique have to be right to produce the goods. It's easy to make a handsome MAG weld, but with this process beauty doesn't necessarily mean strength. It's a shortcoming that must always be in the back of the mind: faulty MAG technique and asking too much of the set will both cause problems, even though the weld's surface looks fine.

So which operator-dependent variables can affect weld strength? The traditional 'stick welding' operator (i.e. someone experienced with MMA) has four things under his direct control: current, arc length, speed of travel and rod angle. If all are handled properly, the result of his efforts will be a sound weld.

The MAG operator must deal with some of these plus a couple of others. He can alter current, voltage, wire extension (stickout) and gun angle, along with both the speed and direction of travel.

One of MAG's selling points is that it's easy to make a reasonable job with very little practice. And it is, because the very thing that most people find hardest with MMA—holding arc length steady—is controlled by the set.

But for consistent, strong welds all the various variables have to be right. Thus it might be a good idea to take each in turn and see what's needed, starting with the one that isn't all that variable.

Arc Length

The more volts and less wire speed, the longer the arc. But that's it; as explained on page 25, the operator can't adjust arc length manually during work because the set's power supply holds it relatively constant. Dialling in a higher voltage lengthens the arc and increases total heat input to the work.

Current

At any given voltage, altering wire feed speed changes current (page 26). But not by all that much, so for large changes in penetration and fusion it's necessary to alter voltage and optimise wire speed as detailed previously. For best results the operator must fine-tune feed speed according to conditions.

Stickout or extension

This one's quite simple. In general, stickout should be kept short,

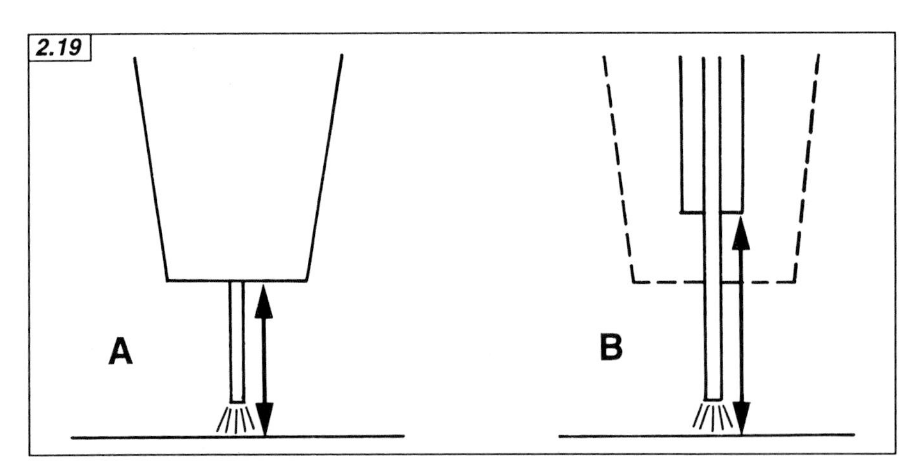

2.19 The two faces of wire stick-out, or extension. **Visual** ***stick-out is what's actually seen by the operator—the length of filler wire extending below the nozzle (A). The power supply, though, senses the length of resistance-creating filler wire between the contact tip and arc (arrow, B). The more recessed the contact tip in the nozzle and/or the greater visual stick-out used during work, the greater the resistance. Welding current drops in proportion, as more energy goes to heating up the high-resistance narrow wire.***

thus minimising electrical resistance downstream of the contact tip and maximising welding current. In practice use 6-10mm visible extension, depending on welding voltage (Figs 2.18 and 2.19).

Where filler wire wants to fuse with the contact tip, try more stickout. If this doesn't help, either the wire speed is too low—or uneven—or voltage is too high for the speed.

It's the other extreme (too much stickout) that's really dodgy. Welding with too much wire beyond the contact tip really hits weld strength.

To see how, try this. With the arc running normally, gradually raise the gun. As stickout grows the arc gets weaker, until finally the filler wire just glows red and won't fuse. By lengthening stickout you've progressively increased electrical resistance between the contact tip and arc, so energy is diverted into heating the wire.

Eventually there's not enough left to run an arc. If that's not bad enough the arc is moved further and further away from its protective gas shield, adding porosity to the weld's woes (see Fig. 2.20). (NB: MMA electrodes don't suffer from this problem as they're much fatter. Consequently they can carry current with less resistance heating.)

Sometimes it can be very tempting to use extra stickout to get round a problem, as when trying to weld in a deep, narrow groove or somewhere that you (or the gun) can't quite reach.

But don't do it. Though the joint may appear to be welded, reduced current means that it's likely to be only stuck together (Fig 2.29, page 38).

Within stickout's upper and lower limits it's acceptable to vary current by varying extension, but only with care. So on a very thin-section butt joint where burn-through looks imminent, a little more stickout can reduce current and prevent disaster. But don't use stickout changes as a substitute for appropriate voltage and wire speed selection.

Speed of travel

Also watch speed of travel, for it causes some of the greatest problems with MAG work.

Going too slow is as bad as going too fast. The trouble is that there's no ready yardstick by which to judge. To do it properly, speed must always be related to position and conditions.

Too-fast travel leaves a stringy bead with reduced fusion and penetration (Fig 2.21). That's fairly easy to spot.

But going too slowly is harder to see, and if anything it is worse for strength. There's a very high risk of poor fusion and penetration, even

2.20 What happens if stickout gets too long? Each weld above was run from right to left, starting and finishing with normal stickout. For the section between the arrows, stickout was doubled. Current dropped and the gas shield was left behind: fusion was hit, bead profile suffered and porosity and spatter grew. The conclusion? It doesn't take much to turn a good MAG weld into a weak one.

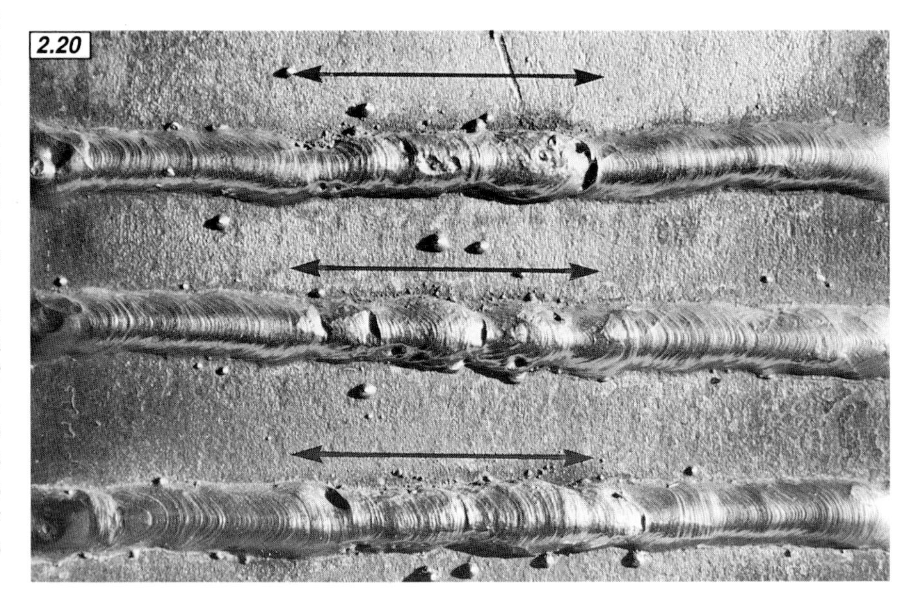

2.21 The Three Bears? Travelling too fast (top) leaves a thin, stringy bead. Travelling too slowly leaves a heavy bead with the risk of poor fusion underneath (bottom), while travelling at a reasonable speed (a little faster than the equivalent MMA run) is just right (centre).

though the weld looks good.

Why? Where filler metal builds up, the wire arcs on to a fat bead of molten metal as it rolls along on top of the intended weld zone. Heat doesn't reach the root and sidewall, so fusion suffers.

Here's a guide to travel speed. Once wire speed has been adjusted to give good welding conditions, travel at a forward speed that deposits a relatively flat weld with good fusion. Watch the departing new-laid bead for height, and the joint's floor, walls and edges for fusion. Work at a speed that keeps the wire arcing towards the front of a slightly sunken molten pool.

Direction of travel

With MAG there's a choice—to pull or to push? Normally the gun is pushed in the direction of travel, with the nozzle making a 70-80° angle to the work.

This is leading angle operation (Fig 2.22A), and it has a couple of advantages. If travel speed is right, wire feeds directly into the hot molten weld pool and fusion is maximised. And compared to MMA, the operator gets a very good view of what's happening.

Alternatively, trailing angle operation is possible (Fig 2.22B). Here, the gun is pulled in the direction of travel. Although this feels more comfortable to someone making the transition from MMA, it's not recommended for most work.

The wire is forced to chase a retreating weld pool and penetration suffers. The arc tends to be less stable, and the chances of weld-weakening fusion 'misses' grow.

Trailing angle operation is only used routinely on vertical down work, because clearly there's no other option! Here it's vital to be aware of potentially poor penetration; more of this on page 40. But the technique has its uses, like an additional ruse to reduce penetration where burn-through can't be avoided in any other way, or to control penetration in the gapped root run of a butt.

Compared with MMA, MAG is more tolerant of angle variation during work. But watch out for using too shallow an angle (Fig 2.23). This not only reduces current through increased stickout, but also tends to carry the arc away from its gas shield.

A couple of comments on visibility and masks. When welding thin sheet there's not much arc light about and it can be hard to see where you're

2.22 Normal practice for MAG work is leading angle operation, i.e. pushing the gun in the direction of travel. This gives the wire a clear feed to the exposed weld pool and the operator a good view of what's happening (A). In trailing angle operation (B), the gun is pulled.

2.23 ***If gun angle gets too shallow, stickout increases (arrow). Current and fusion drop, and the arc tends to edge out of the shield gas envelope.***

going. Changing the mask filter to a relatively transparent grade 9EW helps, but be very sure to swap this for the appropriate item when using currents in excess of 40A—that's equivalent to 15–16V with 0.6mm wire.

If the bank can stand it, keep two clearly marked masks: one for low-voltage MAG work, the other with an 11EW filter for higher voltage MAG and MMA use. The ultimate is an adjustable-density, automatically darkening filter.

MAG—Getting Down to It

After all the preamble, time to do some work!

We'll start right next to godliness. Cleanliness in MAG work is much more critical than with MMA, as MAG wire is very short on contamination-beating ingredients. If it's true that time spent on preparation before MMA welding is never wasted, then it's doubly true for MAG. The process can handle light surface rusting and mill scale, but that's all.

The safest bet is to grind everything back to bright metal—something very often easier said than done. Where possible, get rid of all paint, significant rust, grease, mud and plating from the weld area. Where a grinder won't reach, try flame cleaning and a wire brush.

Most work can be prepared reasonably. But if joint strength is at all critical and you can't make a decent job of removing crud, recognise that MAG, however convenient, has its limitations.

In such cases go back to MMA and hope a lot. Contamination is contamination, whatever the welding process!

Where preparation was skimpy or the parent metal held contaminants in its structure, the arc will sound harsh and the weld pool tend to 'boil'. Porosity will appear along the weld. The only thing to do is to grind out the affected section and reweld.

You may notice a little porosity close to a starting point which quickly clears up by itself. If it's not contamination, it's a temporary shortage of shield gas. When the gun hasn't been used for several minutes, the supply pipework can empty. Giving the trigger a quick squeeze before starting both refills the line and provides a little clean wire for snipping back.

Tacks

If you travel too slowly with MAG, there's a big risk that incoming filler wire arcs only on a built-up pool of previously deposited filler, rather than fusing with the underlying plate (page 34).

So quick blob-tacks are not much use for anything other than thin sheet. What's needed is a definite mini-weld—something 10mm or more long. Blobs offer very little fusion, won't allow much (if any) realignment of parts before breaking and may well shear under heating/contraction stress (Fig 2.24).

Where possible, keep the intended weld line clear by tacking on the

2.24 ***Tiny blob-tacks (A), particularly in anything other than sheet steel, are unlikely to be much good as there will be poor fusion in one or both parent plates. This leaves little leeway for realignment (i.e. hitting with a hammer) and provides a weak spot in the welded joint. Tacks at least 10mm long are much better (B).***

reverse side of a joint. If you're stitch-welding (page 55) position each tack so it forms the start or end of a stitch.

Starts and stops

You're after the cleanest, fastest start possible. That takes high current and good electrical contact, so be sure the return lead connection is sound.

Prepare the ground by snipping off old filler wire at an acute angle. The clean, fine point left behind ensures high current density and promotes fast starting (Fig 2.25).

Maximum current comes from minimum stickout, so snip the wire as close to the contact tip as you can and keep the gun upright when starting. It's a bit of a fine balance—the less initial stickout the cleaner the start, but the greater the chance of filler fusing with the contact tip as soon as the trigger is pulled. That's especially likely at high voltages.

On non-critical work you can get away with leaving the burnt bulb-end of filler wire (Fig 2.26), but always chop it off where high strength is important. An area of poor fusion left at the start of a weld concentrates stress wonderfully, making a prime site for failure.

Note: More complex sets may have burn-back control, allowing adjustment of residual stick-out after welding. Most single-phase sets just let a combination of spool drag and feed motor braking set post-welding stick-out.

All but the cheapest MAG sets have a non-live torch, i.e. no current or wire flows until the trigger is pressed. Thus it's tempting to begin a run by physically touching the filler wire to the start point and then pulling the trigger. The idea works, but it's not really good practice. As the filler wire 'prop' collapses, the contact tip can touch the work and wire will want to burn back and fuse with it. This both damages the tip and delays progress while the mess is sorted out.

Some people deal with excess stick-out before starting by putting a 90° bend in the wire. This may be convenient but it's counter-productive. Not only is arc establishment slower because current is not concentrated into a point area, but a nasty-looking 'whisker' is usually left behind.

The best way to start is by poising the gun over the target and then squeezing the trigger. Good control comes from using two hands: one to steady the gun and the other to direct it and work the trigger. Lay the non-trigger hand over the gun's swan neck (Fig 2.27) for maximum control. Keep the gun upright to minimise stick-out and maximise current; then bring it smartly to

2.25

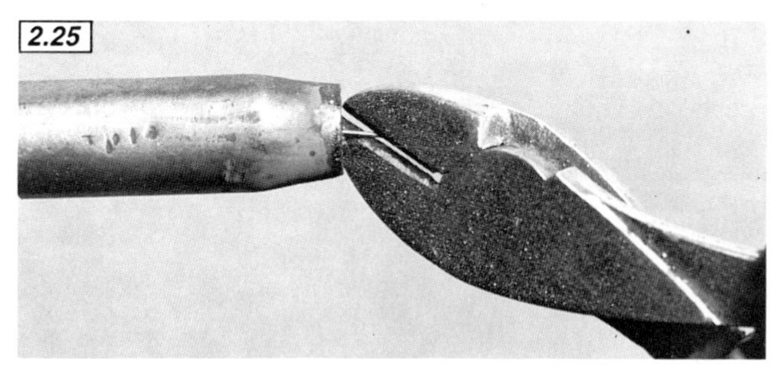

2.25 For fast, clean starts and restarts, snip filler wire off at an oblique angle as close to the contact tip as possible.

2.26

2.26 In contrast to Fig 2.25, over-long stickout and a burnt bulb-end on the filler wire will both delay arc establishment and hurt fusion at the start of a weld run.

2.27A

2.27B

2.27 Gun control is better where two hands are used. Lay one over the torch swan neck (A) or rest swan neck between the first and second fingers (B).

working angle as you move off.

No matter what precautions are taken, MAG starts are prone to lack of fusion. It's a fast-moving process, so you can't wait around and weave while heat builds up as you can with MMA. If you try, the arc is likely to play on a molten lump of filler and fusion will suffer anyway. If you want a really well-fused beginning-of-joint area, use a run-on plate—a piece of scrap alongside the start area on which the arc is first established.

Restarts along an existing weld can also suffer from initial lack of fusion. Even if good they always tend to look a bit lumpy, as it's harder to make an undetectable restart with MAG.

To make the best job, use the MMA technique of starting just beyond the existing run's end, backtracking around the crater and continuing on down the joint (page 11). Where appearances are important, taper-grinding the start area to a hollow pad helps.

On the plus side, building-up the end of a weld run is easier with MAG. Just stop, circle and fill the finish area. Sometimes trapped gas erupts like a tiny volcano during or just after filling, threatening to leave a crack-promoting crater (Fig 2.28).

This often happens when welding old pipe or box section. Coating fume or water inside turns to gas or steam, whose escape route is finally blocked by the weld finish. Alternatively, the crater can come from contaminants left behind after skimpy preparation. Whatever the cause, fill the crater by a short burst of weld immediately the eruption stops, or grind out the defect and reweld.

Watch out!

As with MMA, different types of joint call for different techniques. Knowing MAG's limitations is important, so individual joint types and positional work will be covered in turn. Thicknesses quoted are a guide only.

One big caution applies to *smaller-capacity sets*, and it's this. *There is an upper limit to the thickness that should be welded using dip transfer and 0.6mm or 0.8mm wire.* That limit is about 6mm for single-run joints, or any section over this where multi-runs are needed to fill Vee-preparation.

Note the disaster-promoting combination—dip transfer, thin filler wire and thick sections. Users of smaller farm sets (up to approximately 180A output) are very likely to come across it.

What's the problem? It's that old enemy again—a potential lack of fusion. A thick section acts as a very efficient heat sink, cooling the weld pool. You can compensate for this with MMA by slowing down a little, but not with MAG. All that happens if you try is that incoming wire arcs onto a just-deposited surface ball (page 34).

So you have to keep pushing on. But then there's a good chance that energy input won't be enough for proper fusion. The joint may look welded, but fail catastrophically under load—see Fig 2.29.

Fortunately there are several ways round the problem. If the set allows, switch to 1mm or 1.2mm wire and achieve fusion via higher voltage and thus higher energy input. Or, if the job can be welded in the flat position, stay with the original wire size and shift to spray transfer—its greater heat release gives better fusion.

The third option is also position-related. If the job can't be swung into the flat, there's a good chance that it can be welded vertically upward.

2.28 Escaping steam or contamination-derived gas can crater a good finish, or heave weld metal into a solidified 'pipe'. If the immediately surrounding metal is pinholed too, grind and refill for a high-strength job.

2.29 The combination of dip transfer, thin filler wire and thick plate can lead to trouble. What's wrong with this fillet? Looking at the bead surface in (A), not much. But when stressed, the joint failed suddenly under light load—see how little fusion there was in the vertical plate wall (B).

Working thus slows travel down and lets heat build up. This counter-balances heat loss to the plates, so the original wire can still be used. Use triangular, zigzag or inverted C-shaped weaves, and don't rush—see page 40.

Of course, none of the above ruses might be practicable, but that's rare. Where they're not, switch to MMA.

Joint efforts

First, the ubiquitous **butt joint**. If no edge preparation is given, penetration can vary from full depth to very little, depending on material thickness. Try it and see!

With square-edged (unprepared) sheet that's welded flat with no penetration or root gap, the upper thickness limit for full penetration is about 2mm. Up to 3mm sheet can be handled by leaving a small gap, while vertical down work on anything but very thin sheet demands a gap to help penetration.

Material over 3mm thick will need edge preparation. A single Vee is OK up to 12–15mm, and a double Vee is best over this. If using the latter, balance out distortion by making one Vee double the depth of the other (a 30%–60% split) and fully welding the small side first (Fig 2.30).

Leaving a small step or root face at the bottom of the Vee isn't necessary with MAG, as root penetration can be more readily controlled than with MMA. If necessary, a small weave can be used on the root run to control penetration.

Horizontal/vertical joints are those where you're working side-to-side across a vertical plate. On-the-flat rules apply to butts, modified only by biasing the gun angle slightly towards the top plate to hold weld metal against gravity's pull. For horizontal/vertical fillets, see next.

Compared to butts, **T joints** offer a greater heat sink. You're working on the edge of one plate but well away from the edge of the other, so in total there's more metal to absorb the heat. Higher energy input is called for; one voltage step up over the equivalent-thickness flat butt is normally enough. Balance wire speed appropriately.

Multi-runs are fine for both flat and horizontal/vertical fillets, bearing in mind the risks of using small filler wires and dip transfer on thick sections as discussed above.

If risks look significant, the remedies suggested apply again. But be careful with spray transfer; ideally weld a fillet in the flat position so it forms a Vee under the gun. Spray transfer's pool is hot and fluid, and h/v fillets can suffer from weld sagging and undercut.

Sequence multi-run fillet passes as you would for MMA. That's a root run first, then a half-overlap on the bottom plate and half-overlap on the top and so on (Fig 2.31). Lower the gun angle slightly to fire weld metal up on the vertical plate, and watch for undercut.

Don't slow down to fill a fillet, or poor fusion will likely be the result. Try to avoid weaving for two reasons—it's hard to keep fillet leg

2.30 Thick-section steel can be welded with MAG as long as sufficient heat is available—see page 38. For material over 15mm thick, use asymmetrical double-Vee preparation split 30%/60% as here. Weld the smaller Vee first, completing all runs before tackling the bigger part. With this technique, distortion is largely self-correcting.

2.31 Horizontal/vertical fillet multi-runs (A) are sequenced the same as in MMA work, i.e. a root run, then a run half-overlapping the lower plate and another half-overlapping the vertical plate, and so on (see page 14). Many passes went into this h/v fillet in 25mm steel (B), with no allowance (deliberately) made for distortion—see what the cooling weld metal has done for vertical plate alignment.

lengths equal, and greater heat input makes undercut more likely.

Lap joints are only a different form of fillet, so the same rules apply. Use a slightly higher voltage than for the equivalent-thickness flat butt, don't work too slowly and avoid over-burning the top edge.

Vertical work

Vertical work is generally easier than with MMA, as there's no slag to complicate matters.

General points first. The weld pool is easier to control if current is kept low, so work in the lower end of the 'good welding condition' wire speed range—see page 31.

Gun angles are much the same as with MMA, but are less critical. Start with the nozzle just below the horizontal when viewed from the side, and centred over the joint line when viewed from the front. Angles can be adjusted to help control fusion and penetration—dropping the gun increases stickout, lowers current and makes the pool more controllable. But don't overdo it, or fusion will drop away and there's a risk that the arc will move outside the gas shield.

In general, keep the gun centred over the joint line to balance heat input into both plates. The obvious exception here is when welding thin material to thicker stuff, when heat can be biased towards the thick plate for better fusion and control.

Vertical down work has limited application with MAG. Penetration is never better than poor, so the technique is best restricted to sheet under 3mm. On plate thicker than this there will always be a gross lack of both penetration and fusion.

Butt joints will always need a penetration (root) gap unless material is very thin indeed, and vertical down work should not be used for anything other than an unweaved root run or for filling gaps.

Gaps, you say? No shortage of these in agricultural repair work. In such cases vertical down's lack of penetration can be turned to advantage, allowing very wide joints to be filled without burn-through. Use a side-to-side weave to bridge big holes (Fig 2.32).

2.32 Vertical down work with MAG is never strong, thanks to poor penetration and fusion in anything but thin sheet. But it has its uses: this long open corner joint in 12mm plate had a small gap at the top and a 15mm gap at the bottom. It was welded in one pass, first with no weave and then with the gradual introduction of an inverted C-pattern to bridge the increasing gap.

In contrast, **vertical up** welds are potentially very strong. Control is easier than with MMA, so going upwards shouldn't cause fright. Work is slower thanks to the need for weaving, so there's good heat build-up and hence fusion. Penetration is also excellent as molten metal drains back from the weld pool, leaving clear parent plate for the arc to work on.

For fillets, open corners and Vee-filling, a triangular weave is best. Work deliberately, pausing slightly at the outer corners for good infill. Don't move up too far for every traverse; aim for a compact series of weld steps (Figs 2.33 and 2.34). Keep

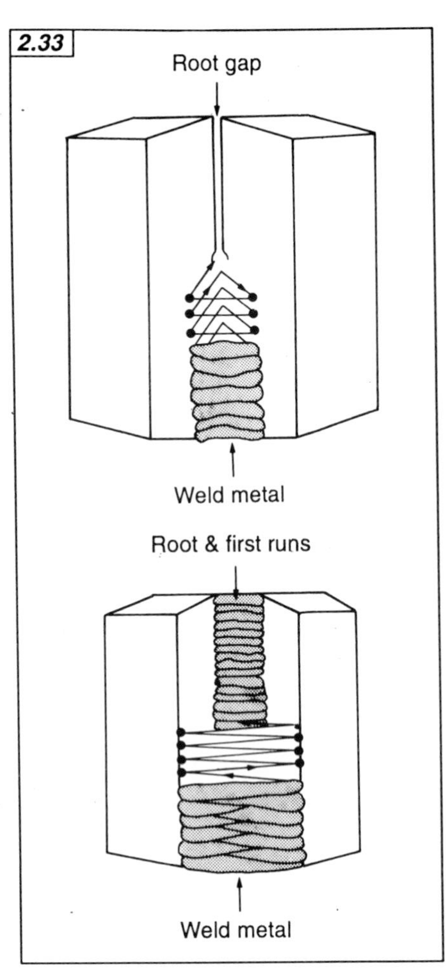

2.33 Good penetration into a Vee comes from a triangular weave (top). The arc is taken into the root with a positive pushing movement: follow the arrows. Move the gun rather than letting stickout grow. The triangle is slanted upwards to give progress. Work deliberately, pausing at the extremities to allow filling (see lines) and not travelling too far upwards for every traverse. In tight Vees, the nozzle can restrict access. If it looks as though you won't be able to reach the joint root without excessive stickout, use MMA. A diagonal up-and-across weave (bottom) can be used on flatter faces, either to cover more width or to give a better-looking finish than an H-pattern movement. An inverted C gives a neat finish, too. Whichever you opt for, don't rush. Work deliberately, pausing at the black circles. Where vertical work on anything other than thin sheet is called for, travel upwards. Done properly, this guarantees penetration and strength.

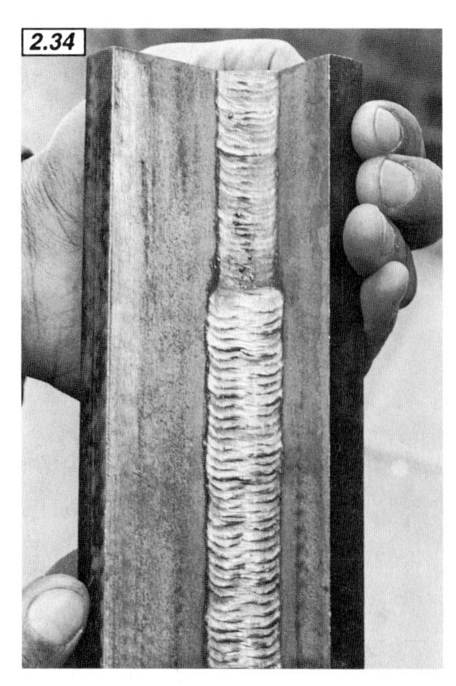

2.34 *Fig 2.33 turned into metal: three vertical-up passes in 12mm plate. The triangular-weave root run can't be seen, but the following two can—and they're examples of good control. The second was another triangular weave, while the third's necessary width came from a diagonal up-and-across movement. Current (hence heat input) was controlled by using relatively low wire speed, i.e. working in the lower part of the 'good welding condition' range—see page 31.*

moving across the weld face, or you'll end up with a weld face that looks like melted candlewax, plus excessive penetration. Subsequent run(s) can be made using a side-to-side weave (Fig 2.33), moving sightly upwards with each traverse and pausing at the edges. Alternatively an inverted C-shaped weave works well, but watch for changes in stickout.

Vertical up's good penetration sets a lower limit to both material thickness and gap size that can be handled. Sheet under 3mm is prone to undercut and eventual burn-through, while working in a gap means difficulty in controlling penetration and maintaining a continuous arc. Wire will shoot through, interrupting the arc and leaving an untidy 'whisker' on the far side.

Finally, overhead work. Some say this isn't recommended with MAG, but it's quite feasible. Use similar settings to equivalent flat welds, but travel a little faster to avoid excess weld metal build-up. Both straight root runs and subsequent weaves are possible. One thing is important—keep an eye on the nozzle, as falling spatter blocks it very quickly, especially at high voltages.

Faults and Remedies

What follows shows common faults and suggests remedies for them, split into sections covering technique, equipment setting, equipment faults and oddities. This final section draws together much of what's gone before, so it refers to illustrations used previously and also includes a few new ones.

Technique-based problems

- **Fault:** Travelling too fast. Weld bead is 'stringy' and irregular; ripples are pulled into Vee-shape. Fusion/penetration is poor. **Remedy:** Slow down! (Fig 2.21, page 35).
- **Fault:** Travelling too slowly. Bead heavy and wide, ripples U-shaped. Risk of poor penetration and fusion. **Remedy:** Speed up! (Fig 2.21, page 35).
- **Fault:** Stickout too long. Bead irregular, fusion patchy, much spatter (Fig 2.20, page 34). **Remedy:** Reduce stickout to 6–10mm working range (Fig 2.18, page 33).
- **Fault:** Weld start poorly fused, or restart not continuous with existing weld metal. **Remedies:** Where fusion is poor, check return lead connection (high resistance lowers current and hence hurts fusion). Snip back old filler wire at acute angle to 6mm stickout before starting. Wait fractionally for arc to establish before moving off, but don't overdo it. If necessary, use run-on plate alongside weld area to establish arc and move on to joint. Fig 2.35 shows poor restart on existing weld, with incomplete overlap of new/old metal. Avoid this by circling back over pre-existing crater to fill it.
- **Fault:** Severe undercut and sagging, much too heavy deposit in vertical up work (Fig 2.36). **Remedy:** Bring heat input down by adjusting voltage, and reduce filler wire

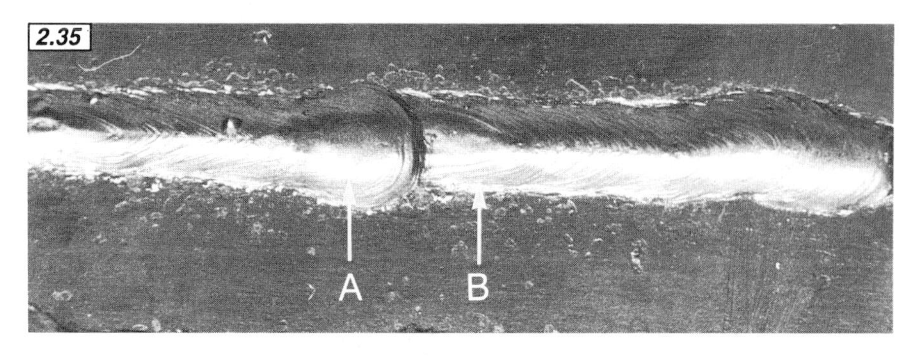

2.35 *A poor restart. New metal (A) hasn't overlapped the old finish (B), leaving a weak spot.*

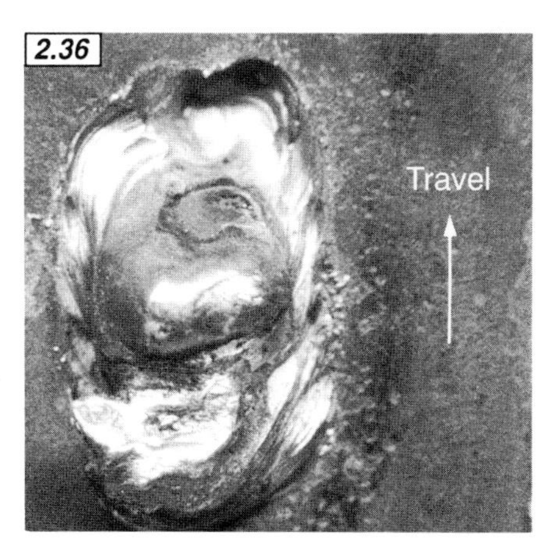

2.36 *Too much heat input and poor technique makes a mess of vertical up work. But see how molten metal drains back from the weld pool, giving great potential for penetration (arrow).*

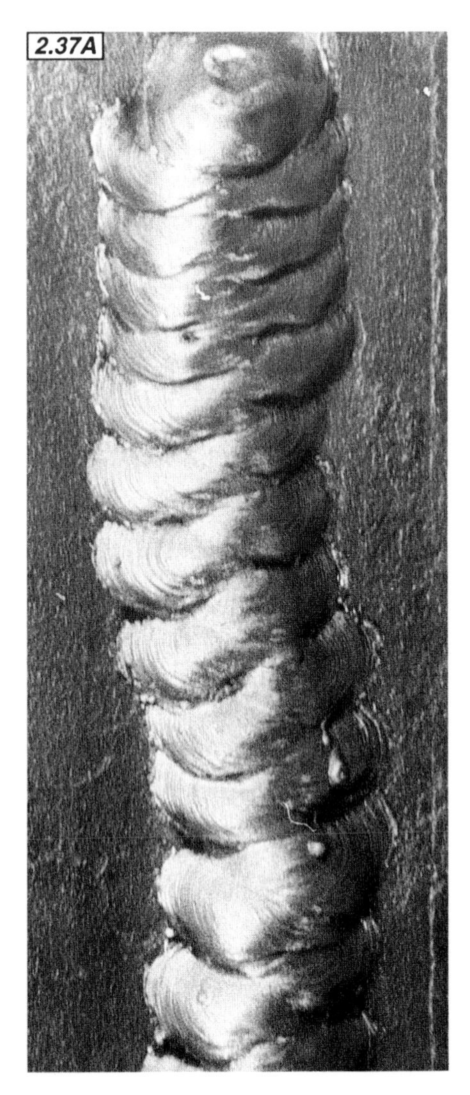

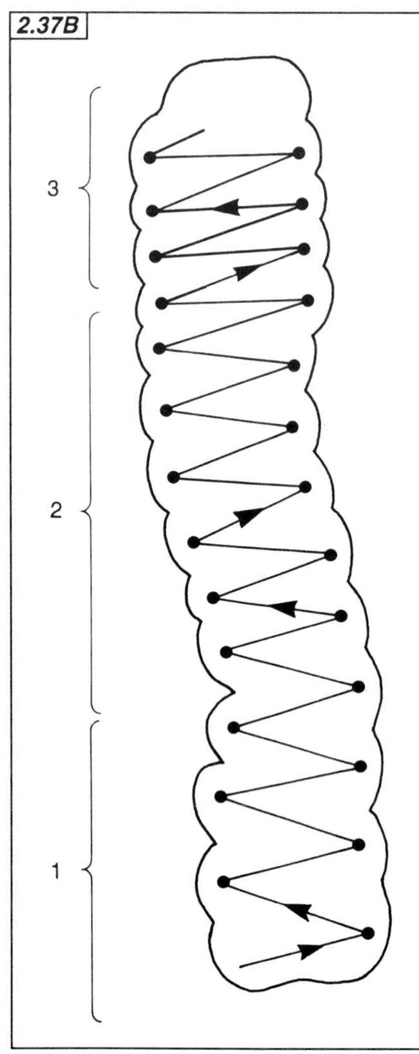

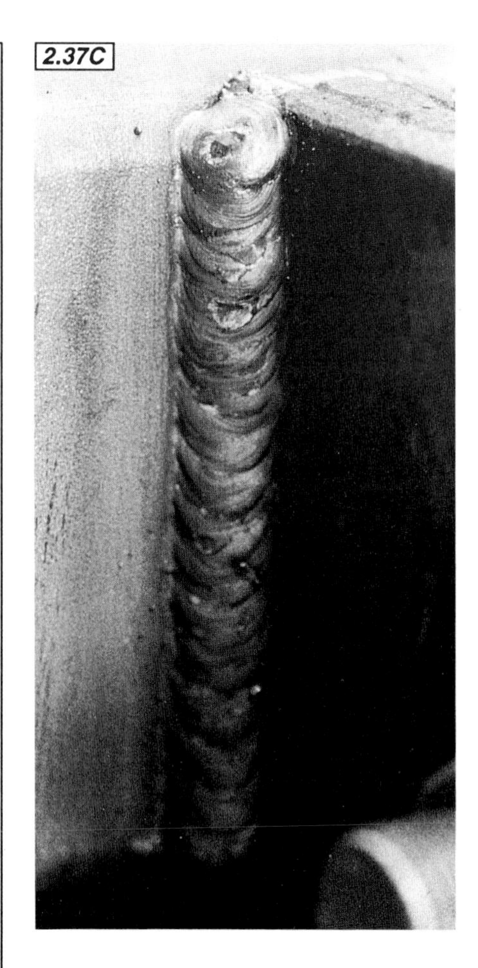

2.37 Common vertical up weave problems (A) and their breakdown (B). Consistent weaving gives a good, strong result (C).

volume by setting wire speed to low end of acceptable range. When weaving, travel smartly across weld face to avoid heavy build-up. Angle each traverse slightly upward to limit metal deposited, but pause at edges to fill joint. Fig 2.37B shows several faults. For a start, the weld wanders off-line. Its lower section (1) has too much upward movement for every traverse, giving weak, big-step appearance and over-heavy metal deposit on face. The middle section (2) is better as upward travel per traverse is reduced. The upper section (3) is about right, with little upward movement per traverse and quick passage across the face. Result: a compact, fast weld with controlled metal deposit.

• **Fault:** Poor joint area preparation. Result: Arc won't run smoothly and develops 'machine-gun' staccato crackle; stubbing (page 26) may occur. Problem may clear itself along joint. Weld quality uneven, probably with areas of pinholing. Joint strength poor. *Cause:* Patchy contamination along weld zone, typical of rust and oil. **Remedy:** Grind out existing poor weld, thoroughly grind unwelded area and try again. Flame clean or wire brush areas where a grinder can't reach. If good preparation is not possible, don't use MAG where high strength is important.

Equipment setting problems

• **Fault:** Voltage too low. Bead is round or oval, thin and humped on plate, with very little fusion visible at edges. Penetration and joint strength both poor (Fig 2.16, page 32). **Remedies:** Increase voltage (optimising wire speed each time) until bead flattens and fuses. If only a small increase in fusion is required, supply more current by increasing wire feed speed. Upper wire speed limit is felt by onset of stubbing. Fine heat input adjustment is possible by decreasing stick-out, thus increasing current. NB: A humped bead can also come from too-high wire speed with correct voltage. To pinpoint this, reduce speed step by step. If bead flattens and fuses, voltage is okay.

• **Fault:** Voltage too high (Fig 2.16). Penetration/fusion excessive, leading to undercut. Burn-through very likely on thin sheet. Bead flat, may be irregular, looks 'burnt' with excessive spatter. Ripples often pulled into Vee-shape as operator subconsciously speeds up travel.

Remedies: Reduce voltage and re-optimise wire speed. If necessary/possible, go down to next wire size. For small heat input changes, reduce current by first reducing wire speed and then increasing stickout to maximum of 10mm. Lower wire speed limit is reached when arc starts to splutter and wire burns off as visible blobs. Where necessary, weave to spread heat.

- **Fault:** Weld porous. In extreme cases the arc sounds harsh and the weld pool 'boils'. Deposited weld metal is coarse, like mousse or Aero chocolate. *Cause 1*: Lack of shield gas, which, when absent altogether, gives rough-looking weld with little fusion. Low flow gives heavy pinholing and possibly weld finish defects (Fig 2.38). **Remedy:** Check that cylinder has gas and spindle valve is open! If okay, look at regulator setting, gas release holes in gun and nozzle for spatter crusting. If all good, follow gas path through set, checking for flow after solenoid valve and before gun. Hole(s) at start of otherwise-good run can result from lag before shield gas flow re-establishes after long period not welding. **Remedy:** Quick pull on trigger before restarting work to refill system. *Cause 2*: Contamination along weld line. Grind and reweld. If good clean-up not possible, use MMA. *Cause 3*: Filler wire rust-covered. Gives weld similar to A and/or B. Clean before feed rollers using spring clip/steel wool over wire, followed by clip/cloth pad (Fig 2.9, page 28). If this can't cope, strip wire off spool until clean coils reached.

- **Fault:** Gun consistently jumps in hand, smooth arc noise breaks up while stickout visibly lengthens/shortens. *Cause*: Stubbing. Wire is being fed faster than it burns back, thus physically pushing gun away from weld area. *Cause*: Wire speed too high for voltage, so readjust.

Equipment-based troubles

- **Fault:** Arc refuses to run steadily. *Cause 1*: Torch guide tube coiled or bent at too-sharp angle, especially where high wire speed needed. Straighten and retry. *Cause 2*: Worn, damaged or loose contact tip—check condition. Discard if next-size-up wire passes through or if correct-size wire won't slip through readily: spatter part-blocking exit hole in tip is common. *Cause 3*: Weld zone contamination. Thoroughly regrind as-yet unwelded area, grind defective weld and redo. *Cause 4*: Wire feed rollers slipping. Check roller pressure, spool drag setting and wire condition. Wire should be rust- and oil-free; spool drag set to just stop overrun when trigger is released, and roller pressure set so wire just stops when braked between fingers. *Cause 5*: Dirty torch liner. Blow out with dry compressed air; replace altogether if necessary. *Cause 6*: Wire drive motor or speed control circuitry unstable. Contact machine supplier for fix.

- **Fault:** No wire feed or wire speed uncontrollable. If motor not running with trigger pressed, check power on(!), speed and voltage selected on front panel, fuses for mains supply and in set. Check gun trigger contacts. If motor operates but no feed, check roller setting and wire spool for tangles. If feed speed uncontrollable, control circuitry has gone berserk—telephone maker.

Oddities

- **Fault:** Joint failure through inadequate fusion (Fig 2.29, page 38). Caused by insufficient heat input: voltage/wire speed too low, forward speed too slow (arc works on rolling ball of metal, rather than biting into parent plate). May happen if welding plate over 6mm thick with 0.6/0.8mm wire and currents under 180A. **Remedies:** Change to thicker wire/higher current if available. Use

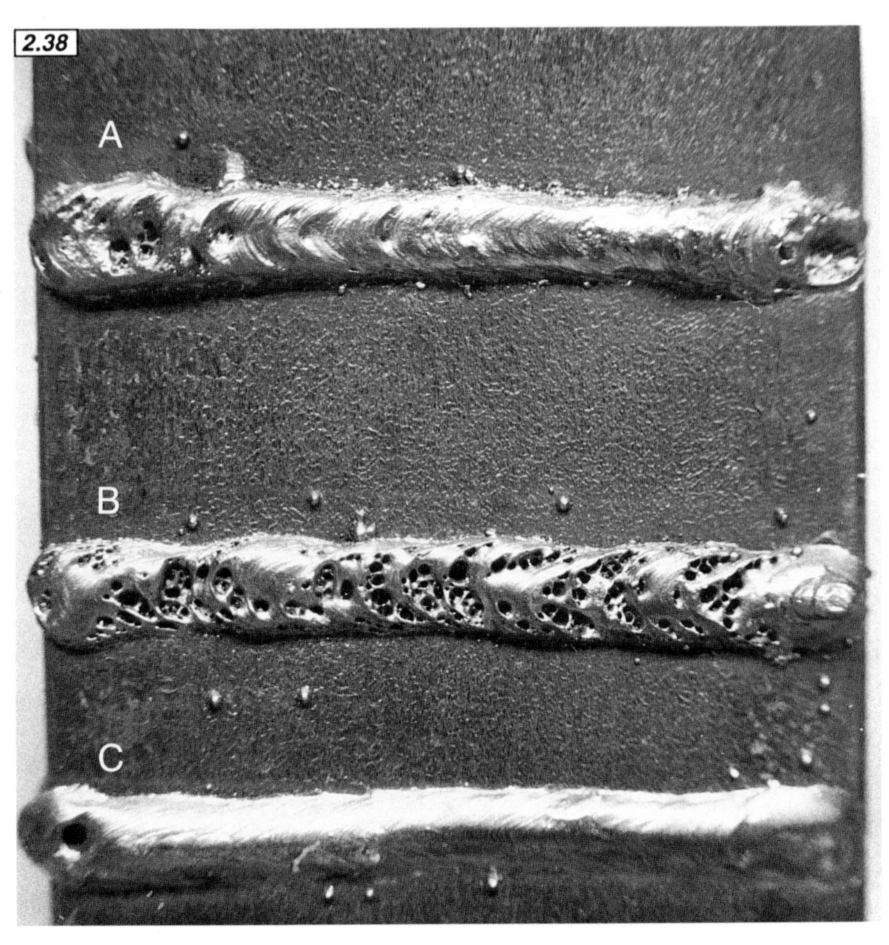

2.38 Shield gas effects. No gas (A) leaves mousse-like bead with some pinholing. Inadequate gas flow (B) gives heavy pinholing; weld zone contamination gives the same effect. Adequate gas flow (C) shields weld and results in good-looking bead.

spray transfer (if set allows) for flat work and horizontal-vertical fillets. Weld vertically upwards in dip transfer mode to increase heat build-up. Travel at speed that keeps wire arcing to front of visibly fusing molten pool. If extra heat input not possible through set output/positional limitations, switch to MMA.

- **Fault:** Volcano-like crater or pipe-like knob at end of run (Fig 2.28, page 38). Snapping off the projection may reveal weak, pinholed weld metal underneath. Caused by escaping gas from contamination or steam. **Remedy:** Fill crater with burst of weld while still hot, or grind off projection and reweld.

Section 3
GAS WELDING AND CUTTING

It's sad, but as arc equipment falls in price and thin-sheet handling MIG/MAG sets creep onto farms, buyers seem less interested in gas welding gear.

Why sad? Because oxyacetylene really is your 'flexible friend'. What else can fusion weld, bronze weld, braze, cut and heat so conveniently? Even if the system's welding ability isn't tapped, having a source of heat for bending, straightening, cutting and unseizing (the latter for bolts and the like) is more than just convenient—in any serious workshop it's essential.

Buying—new or secondhand?

Unless you're quite sure what to look for, buying gas equipment secondhand is probably not a good idea. It's not hard to blow yourself up, and using worn-out gear is an easy way to go about it. So take care over that farm-sale set or local-paper bargain; safety with gas must come first. Look for obvious faults—brutalised fittings, cracked/damaged gauges, damaged seatings or split/cracked hoses. Oxygen cylinders are now being filled to 200 bar (3,000psi) pressure. Check that a secondhand gauge is marked '200 bar service' as a minimum requirement. Today's new regulators are marked '230 bar service' and cover higher filling pressures to come. See Fig 2.2, page 24.

See Fig 3.1 for naming of parts. On the cutting blowpipe (or 'torch'), check the number of pipes between nozzle head and handgrip. BOC's old 'MS' type had only two, and could explode if not properly purged before lighting. When the old types are sent for reconditioning, Murex (previously BOC) swap them for later three-pipe versions (Saffire 3, NM250 and NM400), but there may

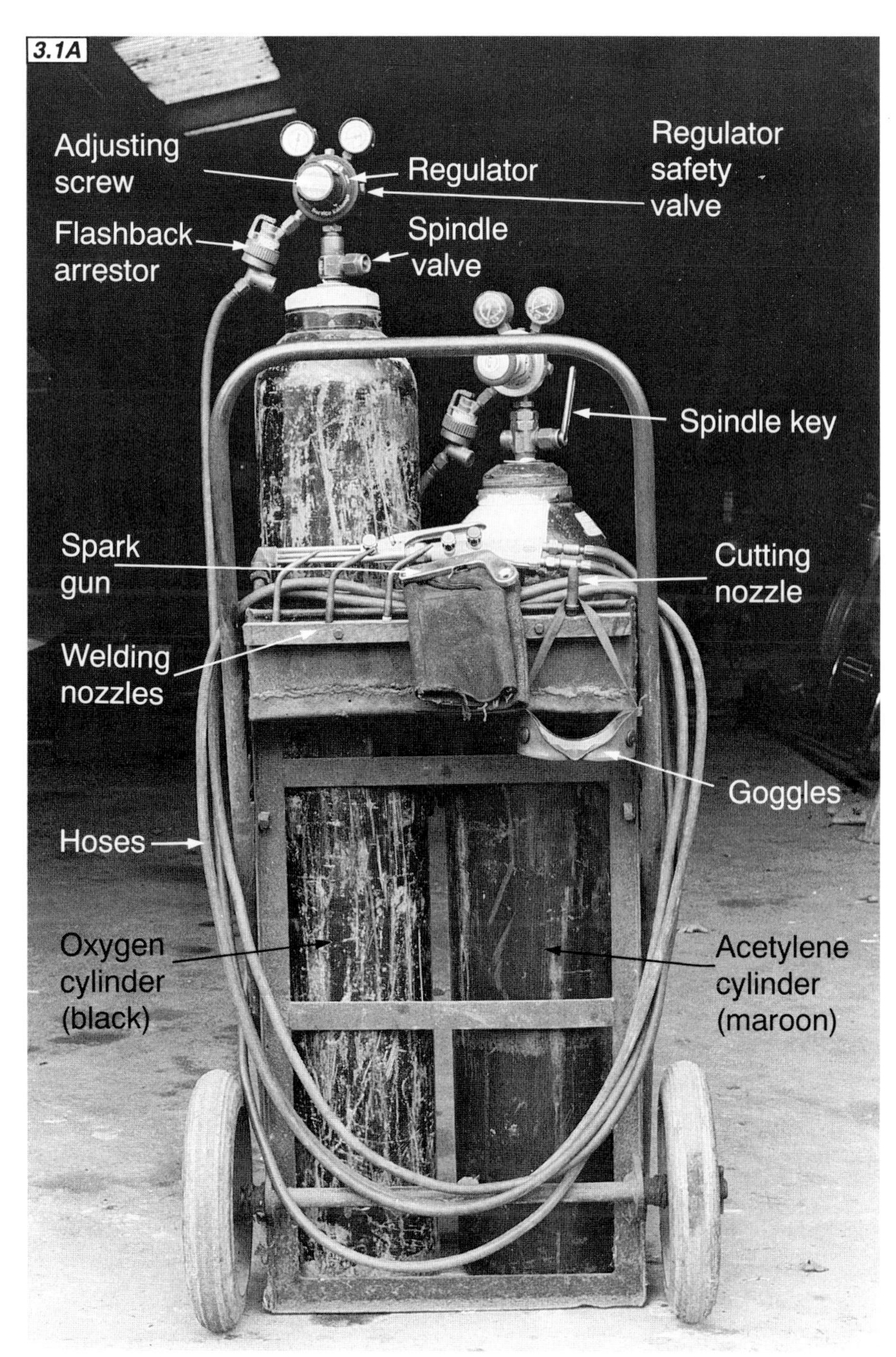

3.1 Equipment required for oxyacetylene welding and cutting (A and B). Note the flashback arrestors—very much advisable for safety. (Continued on next page)

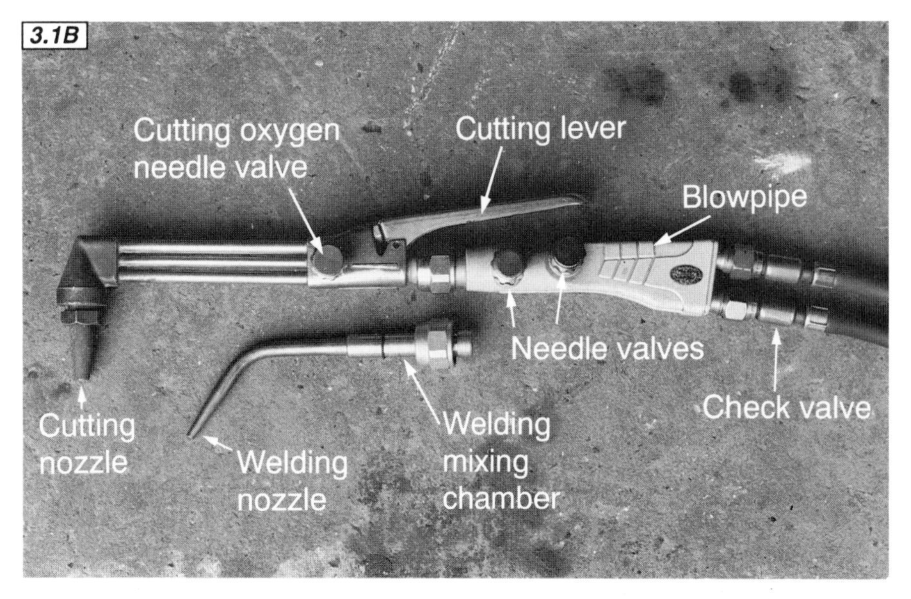

3.1 (continued) Full-size cylinders are heavy brutes to shift; portability comes from smaller cylinders on a trolley (C), which accept normal peripherals. Cylinder life is fine when welding, but short (oxygen especially) when cutting.

still be some about. With any gas equipment the motto has to be *'If in doubt, don't buy'*. Let someone else have the bonfire.

Tread carefully too if you see complete gas outfits for sale. Cylinders remain the property of the supplying company, and won't be the vendor's to sell. Before a sale, old cylinders should go back to the nearest depot and any outstanding rental should be settled. Regulators, pipework, etc. can then be sold, with the buyer renting fresh cylinders in his/her name.

The safest—if not the cheapest—way to own gas gear is to buy new. Looked after, it'll last a long while. Work out your requirement in terms of thickness for welding/cutting, then trundle off to a reputable supplier. He'll fit you up with good-quality kit and be there when service is needed.

The big bang

It's wholly possible to do yourself mortal damage with oxyacetylene equipment, but following basic ground rules will keep you safe. No excuses are offered for devoting space to them.

Don't be tempted to flick through this section. Be sure to understand what's what before trying out new or secondhand gear. If in doubt over any procedure, ask the gas or equipment supplier for advice. Better safe than dead!

Gas cylinders are best treated as potential bombs. A cylinder explosion in a workshop doesn't leave many pieces to be picked up—and there's no second chance. The supplier is required by law to inspect and pressure-test each cylinder regularly; so although a fresh cylinder may have leaked away some of its contents before coming home, it's intrinsically safe. Explosions and fires are thus down to bad practice or equipment neglect. Safety is up to the operator.

So how best to look after the kit and (by implication) yourself? The rules aren't hard, though they might seem many.

Cylinder care is as good a place to start as any. Acetylene gas is only slightly compressible before it goes bang, so it's sold dissolved in acetone. The acetone is held in an inert matrix but isn't fixed by it, so laying a cylinder down lets liquid collect in the neck. This attacks rubber regulator diaphragms and hoses.

Thus cylinders *must* be used upright, and vertical storage and transport is recommended. But of course it's not always practical to move a cylinder without laying it down. To let acetone drain back where it belongs, a cylinder that's been horizontal for more than a few minutes should be left vertical for a couple of hours before use. People do get away without doing this—but acetone gnaws away at rubber over time, unseen. Eventually equipment will fail, always with expensive (and possibly with nasty) consequences.

The exception to the rule is fabricated cylinders (Fig 3.2). These have a different internal matrix, and can be transported in any position.

Fitting regulators to cylinders is straightforward. New cylinders arrive with plastic protectors on their outlets, so these must come off. Regulators and fittings for the common fuel gases—acetylene and propane—have left-hand threads, while oxygen equipment is right-hand threaded. Thus the two can't be mixed up. But don't be tempted to swap fuel gas fittings; though the threads allow it, the gases vary in delivery pressures and permissible flow rates. Using inappropriate bits can affect safety.

With the right regulator to hand, clear the cylinder neck of water and dead spiders by a quick 'snift'. Crack open the spindle valve a little, so gas blasts debris out. Then fit the regulators so that their dial faces can be easily seen and their hose outlets don't interfere with spindle key action. Tighten regulators and hose connections with the spanner supplied; if this has long since gone into a tractor driver's tool box, use a normal-length spanner. If joints leak after reasonable tightening, find out why and rectify rather than trying to overtighten or bodge leaks with tape.

Regulator life will be longer if the central pressure-adjusting screw is backed off after use—not between jobs, but overnight or through the weekend. If left screwed in, the underlying diaphragm and spring will slowly take on a permanent 'set' and accurate adjustment eventually won't be possible. There's a safety angle, too—with the regulator screw backed right off, gas can't escape down the hoses when forgetfulness or a leaky cylinder spindle valve leaves gas in the regulator.

Oxygen regulator gauges carry the warning 'Use no oil.' Ignore this at your peril! High-pressure oxygen meeting oil or grease anywhere in the system (including the blowpipe) can give a tidy explosion. If you must lubricate sticky fittings, use only dry graphite.

Should an oxygen cylinder regulator be blown off for whatever reason (or be sheared off in a fall), it's a bit like letting go of the neck of a huge balloon. Only this balloon is made of steel, and nothing much will stand in its way—not walls, not equipment and certainly not a human body. For this reason, don't be tempted to carry or lift a cylinder by foreloader or forklift unless it's very well secured—and lifting by slippery chains is right out. Keep cylinders (including spares) chained into a trolley or tied back against a wall.

3.2 ***Acetylene cylinders have a squarer outline (left) and an internal construction that allows transport and use in the horizontal position without 'recovery time'.***

More shock horror

Bodging hose connections can lead to trouble through leaks, as can using old, perished hoses. When they're out of use, keep hoses looped in gentle bends: wrapping them tightly accelerates cracking. And don't be tempted to repair a hose with a length of copper pipe, as copper reacts with acetylene to produce unstable, explosive acetylide compounds.

Check hose and blowpipe connections occasionally for tightness. Acetylene's distinctive whiff is easily picked up, so nose round the regulator and pipework—it smells like someone's breath after a particularly significant curry. Never, ever use a match to find an acetylene leak!

Oxygen has no smell, so track down leaks by ear or by using a dilute soap solution. Brush it on and watch for bubbles. Leaks from spindle or needle valves can usually be cured by controlled tightening of the gland nut, but don't overdo it.

Only a very foolish man would carry on working after smelling acetylene. But oxygen can build up in the air unnoticed, and it can be lethal. Though not inflammable, it very actively supports burning. Thus oxygen-impregnated clothing will flare like a torch if a spark lands on it—people have been killed this way. It goes without saying that oxygen must never be used in place of compressed air, either to blow down clothing, clear a bench or power air tools.

Finally, for anything short of very heavy cutting, don't open either cylinder valve more than one turn—less is usually plenty—or it can't be shut quickly in an emergency. And always leave the spindle valve key on the acetylene cylinder valve, for this is the one that must be shut first in case of trouble.

That's a run-through of the main safety points. There are others, which will be covered as they come up.

Backfires, flashbacks and arrestors

It's not unusual to hear occasional 'popping' noises when someone's welding or cutting. At best the pop makes you jump and blows molten weld metal to the four winds: at worst you're on the way to big trouble.

So what's happening? Small-scale bangs are backfires. Gas is igniting uncontrollably, rather than burning steadily at the nozzle. Common causes are:

- The nozzle touches the work, or is temporarily blocked with spatter. Gas flow stops, the flame goes out and then reignites with a bang. Don't try relighting with both gases flowing, or another, possibly bigger bang may result. Shut off acetylene first at the blowpipe needle valve, then oxygen. Purge both gas lines and relight (page 49).
- The nozzle overheats through working in a confined space. The gas mix ignites inside the hot nozzle and explodes. A hot nozzle can be cooled by turning off acetylene and plunging into warm water: cold water can cause cracks. Oxygen flow keeps water out. Check nozzle tightness after as thermal shock can loosen it. For relighting, follow procedure above.
- Gas pressure(s) too low. Very common, this. The gas column moves out of the nozzle too slowly, so the flame burns back down it. In marginal pressure conditions the result will be isolated 'pops' or sometimes a machine-gun string of mini-explosions. In worse cases the flame will disappear down into the blowpipe's mixing chamber and burn there. You'll hear a thin screeching noise and see a stream of black smoke coming from the nozzle. In this case act quickly, or the blowpipe can melt. Shut off oxygen to starve the flame (the only time when oxygen is shut down first) and then the fuel gas. Purge both lines (see page 49) before relighting. To stop the banging, increase pressures a little, keeping them balanced if appropriate (page 50).

The consequences of uncontrolled backfiring can be dire, so don't let it go on. Why? Normally, gas is fed down two hoses into the blowpipe. Where feed pressures are not exactly equal (and they seldom are), gas from the higher pressure line wants to push into the lower pressure one. If allowed, the result would be a potentially explosive mix in one hose which could be triggered by a backfire's flame front.

So the law requires that check valves (Fig 3.3) are fitted at the blowpipe end of each hose. These are just simple one-way valves which close if gas tries to backtrack, preventing mixing in the hoses. Should a backfire occur, its relatively low-pressure, cool flame front is stopped by the check valves.

It's worth testing valves occasionally. With the hose disconnected at both ends, try blowing through it. No air should pass in the regulator direction, but flow should be free towards the blowpipe end.

3.3 Check valves are a legal requirement at the blowpipe end of hoses. Test often by blowing through them—the arrow (top right) shows direction of gas flow.

3.4 Nozzles to extend equipment usefulness. The flame cleaner (top) is good where a diffuse heat source is required, like taking off paint or surface rust from otherwise unreachable areas prior to welding. A gouging nozzle (centre) cuts grooves, also useful prior to welding where a grinder can't reach. For rapid, economical heating prior to unseizing or straightening something, the heating nozzle (bottom) is best. All the above come in different sizes and need a different mixer for the blowpipe.

The bigger bang

Check valves are only a *first line of defence*, and can't handle all eventualities. It's possible in normal use to find a situation of extreme pressure imbalance—as when something steps, falls or drives onto a hose, or when one cylinder runs empty of gas. Then the check valves might not stop all cross-flow, and the low-pressure hose fills with a gas mix.

With a cow or tractor parked on one hose, pressure imbalance at the nozzle makes any backfire much more severe. If the resulting flame front is hot and fast enough it'll jump the check valve(s) and enter the hose(s). The gas mix may explode, the regulator(s) may be blown off and the sharp pressure rise can detonate pressure-sensitive acetylene in the cylinder. Nasty—and there's seldom a second chance from this, a *flashback*.

Thus it makes exceedingly good sense to fit a pair of flashback arrestors (Fig 3.5). These snuff out the flame front before it gets to the regulator; some designs simultaneously trip a valve to stop gas flow. Arrestors are not cheap, but neither is a life.

Good working practice will see that flashbacks don't happen. Keep gas pressures balanced (except in cutting gear, which is designed to handle different pressures), watch hoses around stock and machinery and don't run cylinders right down to their last dregs. And before lighting or relighting a blowpipe, purge both lines to make sure it contains only the gas it should.

Step-by-step

Tables 3.1 and 3.2 give a safe step-by-step procedure for lighting and shutting down gas equipment. Follow it to be sure that gas flow through equipment is under control and that lines contain only one gas.

The fuel gas supply (acetylene) is always dealt with first, as controlling the fuel controls any fire. Purging means ridding a system of mixed gas.

3.5 Simple flashback arrestors (A) snuff out a high-velocity flame front before it can reach a cylinder, using an internal soldered link that melts in flame heat to cut off gas flow. They're once-only-use devices. More expensive designs (B) have pressure-sensitive valving and are resettable via an external arm (arrow).

TABLE 3.1 Safe start-up/shutdown procedure for oxyacetylene welding equipment.

1. Make sure blowpipe needle valves are closed. Back off regulator screws if necessary. Check all connections are tight—especially nozzle/mixer or nozzle/cutting attachment joint at blowpipe.
2. Stand facing regulator dials. These usually have blow-out backs, so facing the dial is safest. Open acetylene cylinder spindle valve *slowly* to a maximum of one turn; slow opening means no pressure shock to regulators. Ditto oxygen spindle valve. Leave key in acetylene spindle valve.
3. Gas is now into both regulators. Twin gauge models will show pressure on one dial.
4. Purging and pressure setting are combined. With the nozzle pointed upwards away from neighbours, cylinder and self, open the blowpipe acetylene needle valve. Adjust pressure regulating screw so gauge shows nominal pressure required by nozzle (Table 3.3). Let gas flow for a few seconds to purge system, then close needle valve. Repeat procedure for oxygen. NB: Pressures can be set accurately only under free-flow conditions. Back pressure from a closed needle valve gives false reading.
5. Open acetylene needle valve a little and light gas. Use spark igniter; matches or a lighter means burnt fingers.
6. Open needle valve further until acetylene flame is 'bright'—smoke at tip is all but gone. If flame separates from nozzle tip, gas speed is too high. Close needle valve until it rejoins.
7. Gently open oxygen needle valve and feed in gas until required flame type is achieved (Fig 3.6).
8. Shut down by closing acetylene needle valve first: flame thus goes out quickly. (Turning off oxygen first would leave a smoky acetylene flame to burn up to—and perhaps inside—the nozzle.)
9. Close oxygen needle valve.
10. Close acetylene cylinder spindle valve, then ditto for oxygen. This traps gas inside cylinders.
11. Drain gas from rest of system by opening blowpipe acetylene needle valve; watch line pressure gauge fall back to zero. Repeat for oxygen.
12. If welding has finished for more than a couple of hours, back off regulator adjusting screws. This prolongs regulator life as well as stopping the progress of any gas leaking past cylinder spindle valve. NB: Cylinder pressure gauge may creep back up over time if spindle valve is leaking. Don't be tempted to jam valve shut by hammering key or using pipe for extra leverage!

TABLE 3.2 Safe start-up/shutdown procedure for oxyacetylene cutting equipment.

1–6 as Table 3.1 above.

7. Open oxygen needle valve fully.
8. Feed in oxygen using cutting head spindle valve until flame is neutral.
9. Operate cutting lever and check flame stays neutral. If not, bring to neutrality while holding down cutting lever, using head (upper) needle valve.
10. Post-work shutdown is the same as Table 3.1. After draining oxygen line of gas, close both oxygen needle valves.

Getting Started

Equipment has been safely set up and the lighting/shut down procedures are understood—but there's still the matter of choosing a nozzle and setting pressures. We'll deal with welding first: cutting is covered on page 60.

Table 3.3 relates material thickness to nozzle size and is a good place to start. But experience will soon show that factors other than thickness have to be taken into account. The aim is to find a nozzle that brings work to heat quickly, yet isn't so big that molten metal in the weld pool is hard to control.

Most farms will use oxyacetylene welding on mild steel sheet up to 3mm thick. Work on thicker stuff is increasingly slow, takes a lot of heat, causes considerable distortion and uses more and more gas. Most jobs will be covered with nozzle sizes 1, 2, 3, 5 and possibly 7, though one-off jobs may call for bigger.

Nozzles are usually stamped with their size (1–90) and the maximum gas volume delivered per hour. So a no. 1 nozzle is also stamped '26', a no. 2, '57' and so on. The smallest nozzles are used where little heat input is needed, while no. 90 roasts chickens at a distance and empties cylinders within minutes.

Assuming thinnish mild steel sheet is to be welded, start with a no. 1 nozzle. Suggestions on how to check if this is right for the job are on page 52.

Flame setting

Depending on the balance of oxygen and acetylene being burnt, a gas flame's nature changes (Fig 3.6).

Different flame types suit different jobs. Most welding jobs need a *neutral* flame; bronze welding/ brazing calls for a slightly *oxidising* flame, while welding aluminium and some hard-facing applications need a gently *carburising* flame.

The different types are easily produced, and are most clearly seen through gas welding goggles.

Set pressures and light the acetylene, opening the blowpipe needle valve until smoke at the flame tip all but disappears. Then feed in some oxygen: watch as the flame turns from yellow to white. Feeding in a little more oxygen resolves the flame into three sections—a blue central cone next to the nozzle, a white 'feather' beyond it and an outer, less distinct envelope.

Adding more oxygen sees the feather shortening. While the feather is visible there's more carbon-rich acetylene in the flame than oxygen to burn it, so the flame is *carburising*. It's also cool. Working with such a flame leaves a weld that's hard but brittle.

As oxygen supply is increased, the feather just merges with the central cone: the cone looks 'hazy'. Oxygen and acetylene are now being burnt in equal volumes, leaving a hotter *neutral* flame. Neither carbon nor oxygen is left over to combine with weld pool metal, so potential weld strength is good.

Feeding in yet more oxygen unbalances the flame again. Now there's an excess of oxygen: the cone gets smaller, harder-edged, more pointed, lighter in colour and takes on a spiteful 'hiss'. Now it's an *oxidising* flame, which—where not wanted—will produce showers of sparks and convert molten metal into weld-weakening oxides.

Play around with flame types so that you can produce each at will. As equipment warms up, a neutral flame tends to become oxidising—so watch your flame during work.

Cheaper single-stage regulators can't hold pressure as constant as two-stage versions, so pressure drift can also unbalance the flame during work. But if the flame changes type with a regular pulsing rhythm, watch out—gas pressure is too low on one of the lines, making a check

TABLE 3.3 Welding nozzle sizes, recommended pressure settings and material thickness weldable. (Courtesy Murex)

See how gas pressures rise with nozzle size but stay balanced. Figures are a guide only, and refer to clean mild steel worked indoors.

Mild steel t'kness			Nozzle	Operating pressure				Gas consumption			
				Acetylene		Oxygen		Acetylene		Oxygen	
mm	in	swg	size	bar	lbf/in²	bar	lbf/in²	l/h	ft³/h	l/h	ft³/h
0.9		20	1	0.14	2	0.14	2	28	1	28	1
1.2		18	2	0.14	2	0.14	2	57	1	57	2
2		14	3	0.14	2	0.14	2	86	3	86	3
2.6		12	5	0.14	2	0.14	2	140	5	140	5
3.2	1/8	10	7	0.14	2	0.14	2	200	7	200	7
4	3/32	8	10	0.21	3	0.21	3	280	10	280	10
5	3/16	6	13	0.28	4	0.28	4	370	13	370	13
6.5	1/4	3	18	0.28	4	0.28	4	520	18	520	18
8.2	5/16	0	25	0.42	6	0.42	6	710	25	710	25
10	3/8	4/0	35	0.63	9	0.63	9	1000	35	1000	35
13	1/2	7/0	45	0.35	5	0.35	5	1300	45	1300	45
19	3/4		55	0.43	6	0.43	6	1600	55	1600	55
23	1		70	0.49	7	0.49	7	2000	70	2000	70
25	1 +		90	0.63	9	0.63	9	2500	90	2500	90

TABLE 3.4 Flame cleaning.

Acetylene fuel gas nozzle Type	Fuel gas pressure		Oxygen pressure		Fuel gas consumption		Oxygen consumption	
	bar	lbf/in²	bar	lbf/in²	l/h	ft³/h	l/h	ft³/h
50mm flat	0.49	7	0.57	8	1050	37	1200	41
100mm flat	0.7	10	0.7	10	2000	70	2200	78
150mm flat	0.85	12	0.85	12	2700	94	3000	104

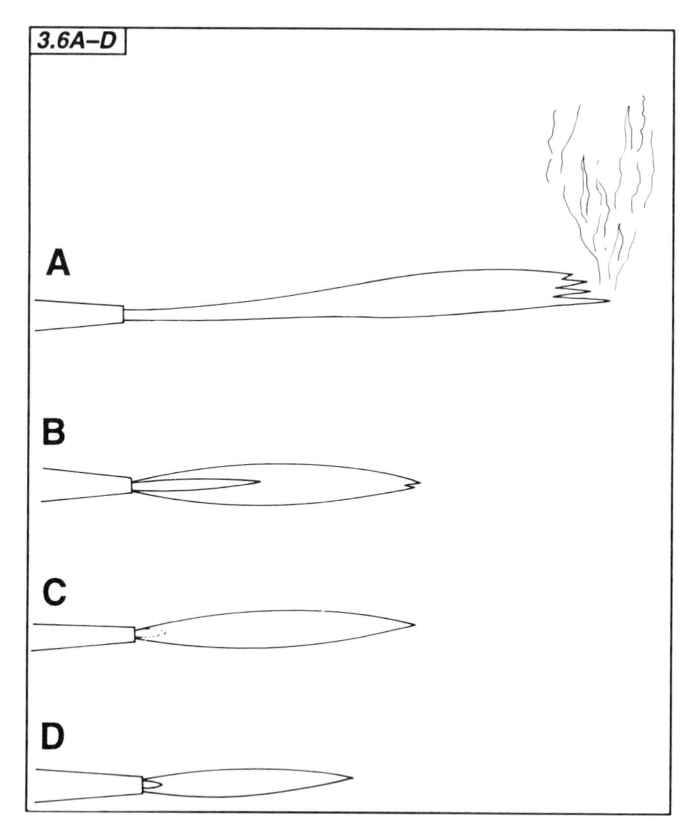

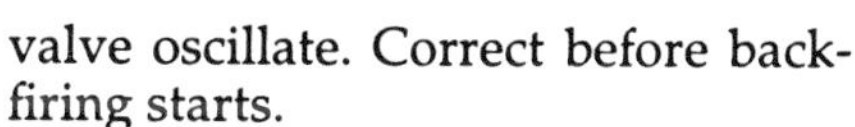

3.6 Getting the right flame is crucial to a good gas weld. With no oxygen, acetylene burns ragged yellow with much soot and smoke at low flow rates (A). If flow rate is too high, the flame separates from the nozzle (E). Feeding in some oxygen quietens the flame and produces a long, white central feather beyond a small central cone (B): this is a carburising flame. Adding more oxygen sees the feather shrinking until it's just merged with the central cone, leaving a faintly fuzzy outline (C). This neutral flame is used for most work. Dialling in more oxygen sees the flame shortening and getting noisier with a tiny, pointed cone (D). There's now an excess of oxygen—an oxidising flame. To check for neutrality, take out oxygen via the needle valve until a feather appears, then add it until neutrality is achieved. Watch the flame all the time during work, for it's prone to change.

valve oscillate. Correct before back-firing starts.

Weld away today

With the smaller nozzles at least, gas welding is a much quieter process than arc's all-action sound and fury. Part-science, part-art, gas welding seems suited to men of easy temperament.

For mild steel work there's a choice of two processes—fusion or bronze welding/brazing. Which to use?

In fusion work, plates are joined by heating until they flow together in a liquid pool. If necessary, filler rod of similar composition is added to build up or fill the joint.

But in bronze welding and brazing, the parent plates are not melted. They're brought to red heat and then a relatively low melting point copper/zinc alloy is flowed on to the surface. This acts like a metallic 'glue', hooking into tiny surface hills and valleys and locking plates together.

To promote cleanliness and surface wetting, bronze work needs a flux. This either comes from a pot or is carried on or in the welding rod (Fig 3.7).

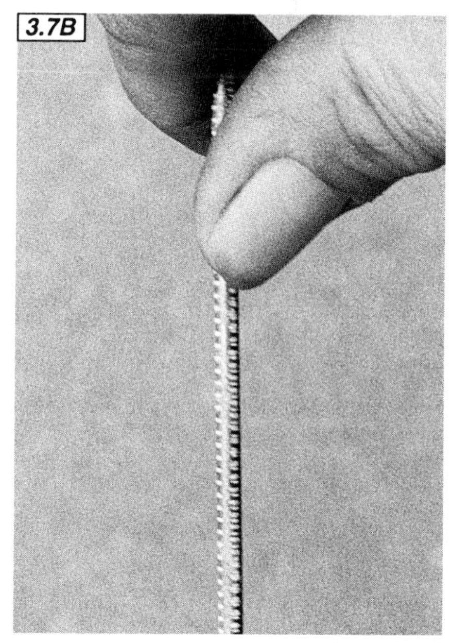

3.7 Bronze welding flux can come in a pot (A) or as part of the rod (B). Here small notches in the rod's flank hold it.

Each process has its advantages, disadvantages and place. Fusion welding needs slightly less critical preparation, and (given good execution) is stronger than a bronze weld. But the process puts more heat into the job, increasing distortion.

Joints for bronze welding need more careful preparation, but the technique's low heat input minimises distortion. Different thickness sections are easily joined, and (given the right rod) dissimilar metals can be welded. And it's easy to do!

Brazing uses the same fillers as bronze welding, but exploits the liquid alloy's ability to move between close-fitting plates or tubes. Capillary action sucks in hot bronze, producing an extremely strong joint with no external build-up (Fig 3.8). Bicycle frame joints are a typical example of this technique's many uses; plumber's Yorkshire fittings use the same idea with solder.

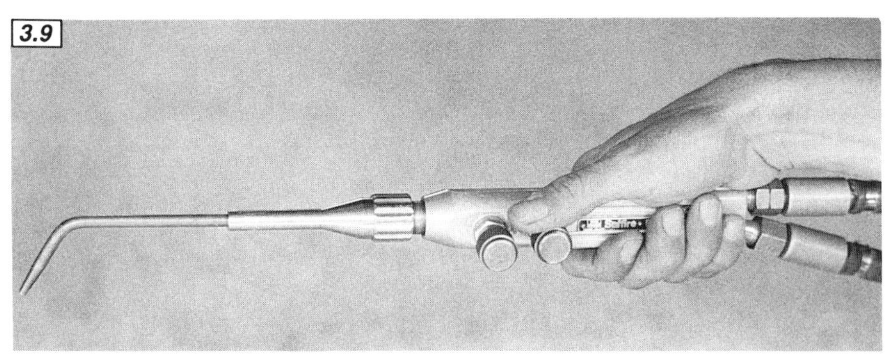

3.9 Before work, rotate the mixing chamber and nozzle so the blowpipe can be held with one thumb over the acetylene needle valve. That way, fuel gas can be quickly shut off in an emergency.

3.8 Brazing relies on liquid bronze's ability to be sucked into a joint by capillary action. See pages 59–60.

Table 3.5 shows common types and sizes of filler rod for fusion and bronze work. Each manufacturer's range differs slightly, but all have the main types represented. Fusion work requires a filler rod of similar material to the parent plates. If in doubt over which rod to use, call your supplier or the rod maker.

Fusion—the practice

We'll look at fusion welding first. For thin mild steel sheet work (up to 1.5mm), a no. 1 nozzle and 1.6mm mild steel filler rod are a good starting point for experiment. Cut a chunk of plate roughly 150mm × 100mm and set it on two lengths of angle iron. This keeps it off the bench, away from the chilling effect of underlying steel.

Before beginning, get the body sorted out. Comfort is essential for good rod/flame control, so lean on any support or (ideally) sit down. Take the time to unwind any twists in the supply hoses, for fighting them is a mug's game. Also rotate the mixing chamber so the nozzle points down at the work: the blowpipe's acetylene needle valve should be under your thumb, ready for quick shut-off in case of backfire (Fig 3.9).

Check your goggles. These should be green-tinted for most work and marked 'GW4'. Use the darker GW5 tint if glare is troublesome.

Drop goggles down and set a neutral flame. Position the nozzle so the hottest part of the flame (around 3mm from the inner cone's tip) plays on the plate (Fig 3.10). If the nozzle is big enough, within a few seconds a clear liquid pool will magically appear under the flame cone.

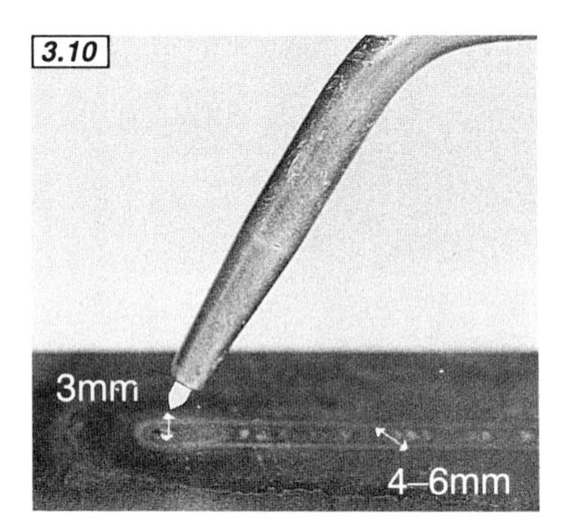

3.10 Maximum heat input comes from working with the flame's inner cone about 3mm above the target. Holding the blowpipe parallel to the work gives the proper nozzle-to-plate angle, thanks to the nozzle's cranked end. Start by making a molten pool on some thin sheet. Push it forward, travelling at a pace that keeps it about 4–6mm wide.

If the pool is part-covered by a shifting, sparkling skin, check the flame is still neutral and that the parent steel is not rust-covered. Flying sparks mean oxides are about.

If a pool is reluctant to appear or flame heat very rapidly burns through the plate, experiment with flame and nozzle sizes as follows.

A range of flame size (and thus heat) is available from any given nozzle. To increase heat, add more acetylene to a balanced flame via the needle valve and rebalance to neutral. The flame is now 'harder', longer and hotter, so will produce a molten weld pool more quickly.

If a too-rapid burn through is

TABLE 3.5 Commonly used gas welding rods for steels.
(Examples from Murex range)

Trade name	*Suited to*	*Diameters (mm)*
Process: Fusion welding		
Saffire mild steel	Low carbon steels	1.6, 2.0, 2.4, 3.2, 5.0
Saffire medium carbon steel	Most low-alloy steels	1.6, 2.4, 3.2, 5.0
Process: Bronze welding and brazing		
Saffire silicon-bronze	Steel, copper, brass	1.6, 2.4, 3.2, 5.0. 6.0
Saffire nickel-bronze	Steel, malleable iron	1.6, 3.2
Saffire manganese-bronze	Cast or malleable iron	3.2, 5.0
Self-fluxing rod for bronze/brazing		
Saffire Fluxobronze K for silicon-bronze	Steel, copper, brass	2.4, 3.2

your problem, produce a cooler and 'softer' flame by taking out a little acetylene and rebalancing.

But there are limits to how far you can go with such fiddling. If you push a hard flame too far, gas velocities get too high. Ultimately the flame separates from the nozzle, and it's likely that the molten pool will be pushed around by high-speed gas. If either happens, go up one nozzle size to bring flow rates back to normal.

Coaxing a smaller, cooler flame from a given nozzle means closing needle valves, thus dropping gas velocities. Again, watch out: below a critical flow rate, backfiring will begin. That'll make you jump and blows the weld pool to the four winds—but worse, it's a potentially lethal precursor to a flashback. *So don't persist with pops and bangs—if you need a cooler flame, go down one nozzle size.*

The aim is to weld with a flame that **you** are controlling, rather than one that's controlling you. So having found a nozzle/flame size that produces a molten pool fairly quickly, experiment by 'pushing' that pool across the plate. Move the nozzle so the pool travels forward, judging travel speed so that it stays constant in width; around 4–6mm is fine. Righthanders work from right to left, keeping the nozzle pointing straight down the line of advance and held at 60–70° to the horizontal (Fig 3.10). Lefthanders work from left to right.

When you're happy with this, try adding filler rod. Snip a fresh 1.6mm mild steel rod in two and bend one end over. This means it won't jab in anyone's eye and lets you spot the hot end when you pick it up!

Balance the rod between spread fingers and thumb of the left hand (Fig 3.11). Position the hand so the rod lies along the line of the weld and makes a 20–30° angle with the plate (Fig 3.12).

Make a molten pool, and then rock the left wrist so the rod tip dips in and out of the pool. Move the blowpipe and rod together across the plate, dipping as you go: travel to maintain pool width at about 6mm (Fig 3.13).

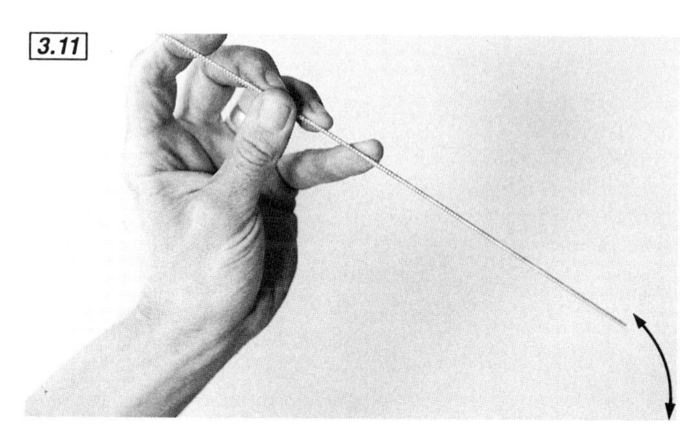

***3.11** Filler rod is best balanced between stretched fingers. Feeding in and rocking the wrist then takes the tip controllably to and from the weld pool (arrow).*

***3.12** Filler rod is held at a much shallower angle than the nozzle, letting it dip cleanly under the latter's tip; around 20° is fine. Keeping it in the flame envelope during work makes sure it's close to melting point and away from oxidising air.*

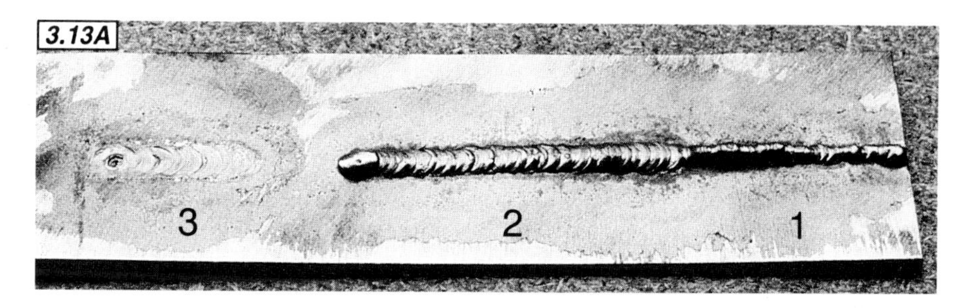

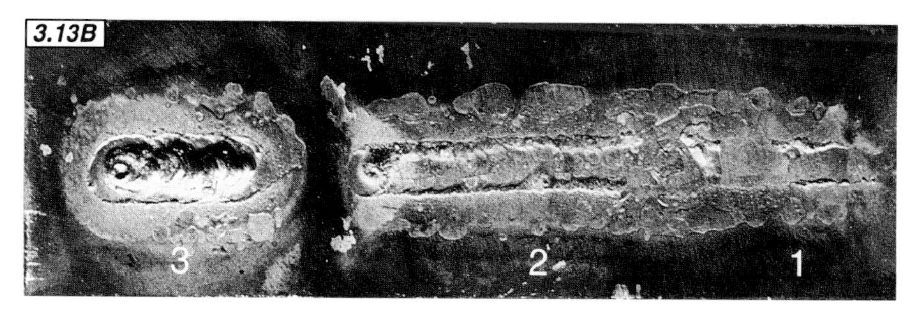

***3.13** Travelling too fast or using too small a nozzle means a narrow weld pool, with little heat available for fusion (A, 1). The resulting bead is thin, humped and not fused into the plate at its edges. Slowing down or increasing flame size lets the weld pool grow to around 6mm wide, improving fusion and bead appearance (A, 2). But too much heat or too slow forward progress results in a wide, sagging weld area, excessive penetration and a fighting chance of burn-through (A, 3). The reverse of the job (B, 1–3) shows corresponding penetration: see how it increases as speed of travel comes down and/or flame size goes up. NB: Example produced by just decreasing speed of travel between 1 and 3.*

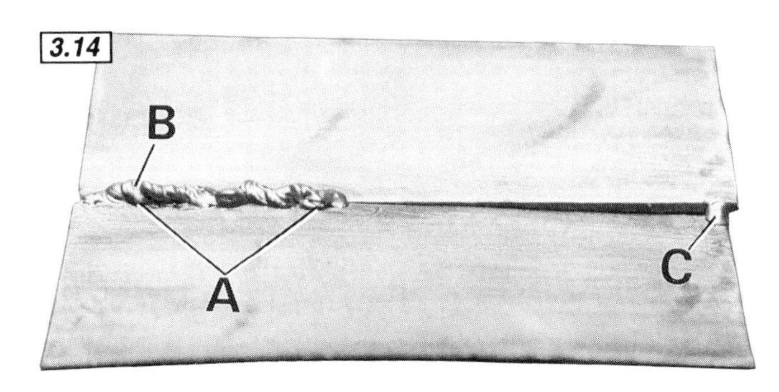

3.14 A small bird flew over and dropped this. The bead constantly changes shape, suggesting uneven rod/flame work. Fusion was poor, as rod was melted by flame heat rather than being added to a molten pool. Where rod solidified on the underlying plate, cold laps are evident (A). Rather than one continuous bead, unsteady progress gave unfused cold shuts (B). Slapdash preparation left the plate edges misaligned (C), wrecking chances of equal fusion in both plates.

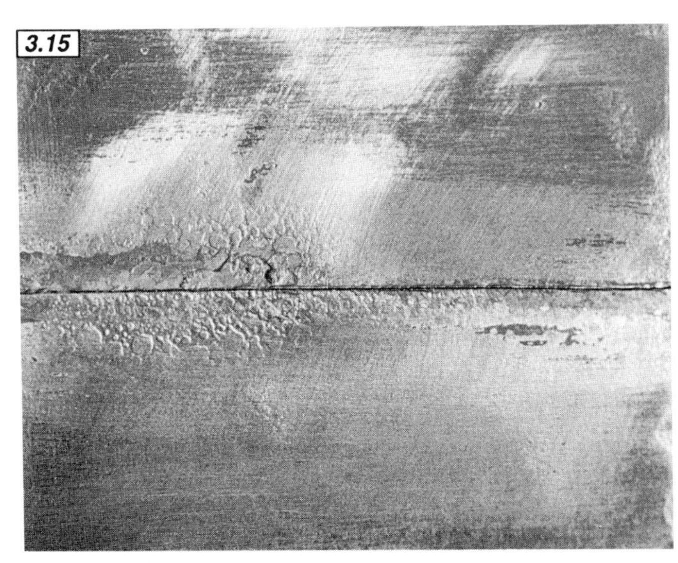

3.15 Turning 3.14 over shows no penetration—quite hopeless.

Keep the rod low and within the flame's protective outer envelope, and slide it in under the flame's inner cone.

Keep the rod rocking in and out of the pool like a nodding donkey. Try to hit the middle of the pool each time; if you catch the front edge or leave the rod in too long, it'll stick. Don't worry—just hold it still until flame heat frees it.

Whatever you do, don't use the flame to melt and drop bits of rod on to the plates. This only produces the renowned 'bird dropping' effect (Figs 3.14 and 3.15), which produces no fusion at all and thus has zero strength.

Good fusion welding is a relatively fast, continuous process in which rod is always added to the centre of a molten pool. But only weld with a flame that you control rather than one that's controlling you. The result of tidy rod/flame work looks like Figs 3.16 and 3.17.

If it's closer to Figs 3.13 and 3.14, don't despair. Use the captions to sort out the reason(s). You may need several tries to get the moving pool/rod act together, but once the penny drops it's not forgotten.

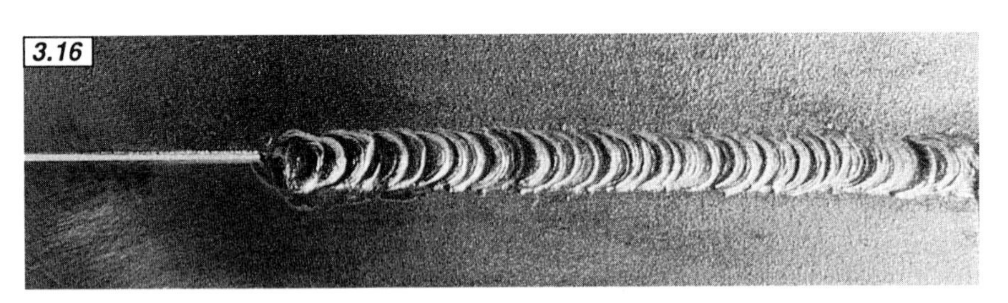

3.16 Steady work with flame and rod leaves an even-width bead with fairly consistent ripples. Edge fusion was good, suggesting that flame heat was about right for plate thickness and desired speed of travel.

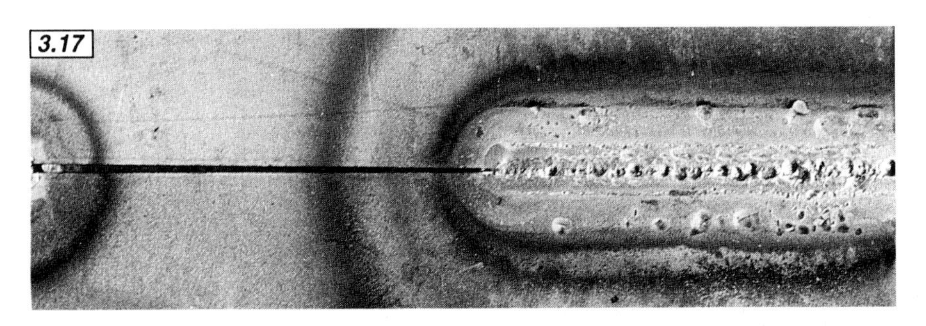

3.17 The reverse of the joint shows pretty consistent penetration, suggesting good preparation (plate edges aligned and tacked, in this case with a small gap) and steady, controlled progress. See how heat blueing is equal in both plates, confirming good flame positioning.

No ifs, just butts

With the basic rod/flame technique cracked, move on to a butt joint. Take two lengths of 1.6mm or 2mm mild steel plate measuring about 150mm × 40mm. Lay them out with a penetration gap equal to (or just greater than) plate thickness.

Bring down the flame and build a tack across one end. Fuse filler metal with one plate, then make a bridge across to the other. Finish by heating the whole tack so it fuses flat into both plates: note that this bit is critical, or the tack will be weak. Regap the plates; they'll have moved through distortion. Tack the centre and other end.

Now weld the joint from right to left, using filler rod. See how the plates want to close together just in front of the flame? That's why a good-sized penetration gap is needed. Nozzle and flame sizes should be such that work is fast but controllable.

If you have to wait around for a pool to form or if it wants to freeze on adding rod, there's not enough heat going into the work for good penetration.

But if holes keep appearing, you're either working too slowly or with too hot a flame. Watch the molten pool very closely: it'll sag just before burn-through, so good penetration is certain if the pool is kept on the point of sagging as it's moved along the joint.

As the joint end approaches you'll have to work quicker. Heat travels up the plates in front of the weld, so the finish area is very hot by the time you get there. Add to this the fact that there's decreasingly little metal left to absorb that heat, and melt-through is a distinct possibility.

If you're not prepared for it, that is. Speed up work as the joint end approaches, and then, at the finish, add filler rod quickly to build up a pad (Fig 3.18). Lift the nozzle straight up when you've done; this lets the weld cool in the flame envelope, which protects it from the air's weakening oxygen and hydrogen.

Flip the finished job over and check penetration. There should be an even bead all along the back (Fig 3.17). If not, either the penetration gap has closed right up or flame heat and/or welding technique wasn't right.

See how the plates have bowed? Distortion is a big problem with gas fusion welding, as relatively slow travel means great heat input. Fast work minimises this, but can't defeat it.

Distortion control

One option here is to use a lower heat technique, like bronze welding (pages 58–60). But this may not be appropriate.

Heat input can be reduced by not welding all the joint. Often it's not necessary, as when fitting car body panels: such sheet work can often be spot-welded. Design a joint so the plates overlap by around 12–15mm, and then drill 4mm holes every 150mm or so in one sheet only. Clamp the sheets together and then fusion weld to 'plug' each hole. But be quite certain that the undersheet is fused, or the job will fall apart in service.

What else can be done to cut down distortion? Where full-length fusion welding is inescapable, frequent tacks help control plate movement. Space them every 50mm or so—much closer than for electric arc work. Make sure each is strong, too, for good tack fusion is essential.

A couple of operational techniques help spread heat around.

Skip welding involves just this. After tacking, divide the joint into 30–50mm sections and mark each with chalk or a soapstone stick (page 63). Then picking and choosing at random, weld one, then another. The idea is to move heat input around so it's not concentrated at any particular area along the joint (Fig 3.19)

The alternative to skip welding is backstepping. Mark out the joint as before, but this time weld the far left-hard section first. Then move to the right, welding one section at a time. Again, localised heat build-up is not so great as it would be with one continuous right–left weld.

Neither of these methods will be 100% successful, but they do help on a long joint. Physical restraint is another option, clamping plates so

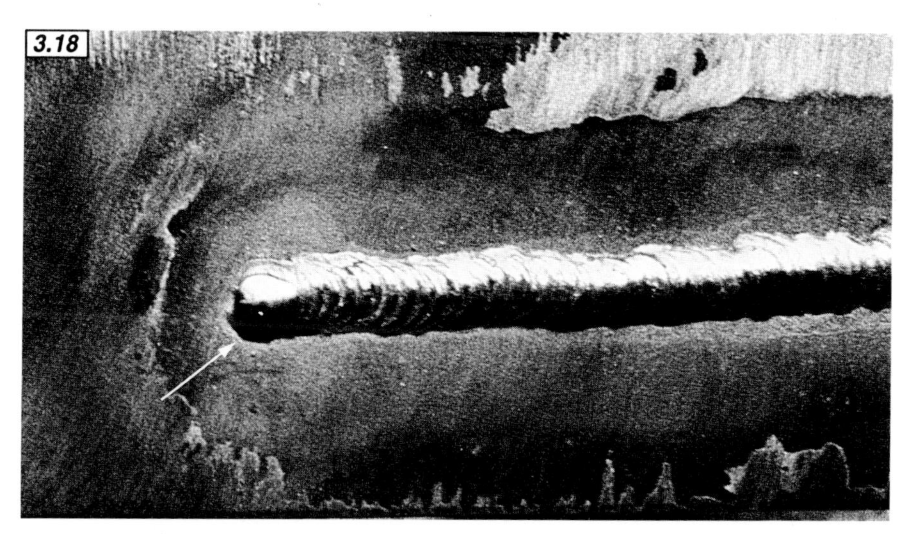

3.18 Building up the finish of a gas weld (arrow) is just as important as with arc; leave it hollow and it's an invitation for cracks to start. Just hold the flame steady and add rod until the crater is filled to bead height.

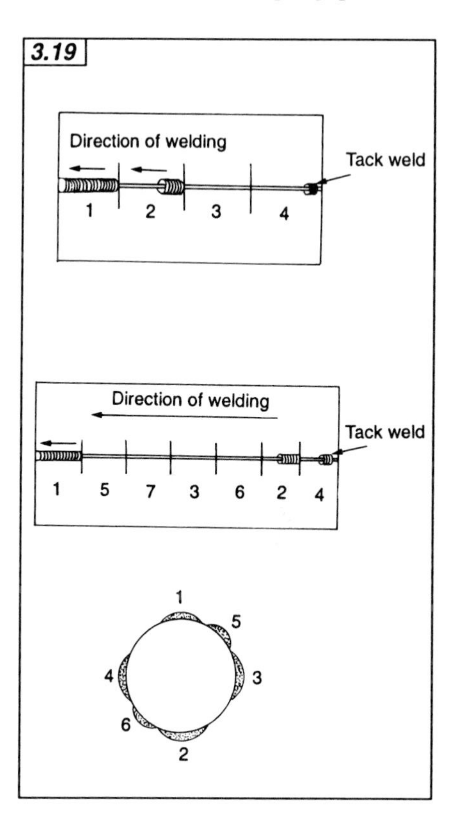

3.19 Backstepping (top) helps minimise distortion, whether with gas or arc work. Sections 1–4 are marked off, and then each welded in turn from right to left. In skip welding, sections 1–7 are welded in random order, each again from right to left. When building up a worn shaft or joining a broken one, lay beads on opposite sides to balance heat input (bottom).

they can't move. Where plate and clamp sizes allow, this works pretty well.

Distortion builds up as work progresses. If allowed to go too far it'll upset unwelded plate alignment, which in turn will reduce the chances of full penetration. Thus on a long joint it's worth stopping to see what's happening ahead, then using agriculture's best distortion-corrector (a large hammer) to set plates straight. Trying to weld buckled, separated plates is irritatingly difficult and certain to give a weak result.

Something easy

Filler rod isn't always needed. If joint design and thickness allow, its edges can be run together into a pool and carried along by flame heat.

Open corner joints in sheet are great for this. Take two bits of 1.6mm or 2mm plate, about 150mm × 40mm. Tack them to form a right-angled roof shape (Fig 3.20), keeping edges parallel and close together. This is vital, for big gaps mean burn-through and misalignment hurts fusion.

Bring the flame down on the right-hand tack and remelt it into a pool. Now move forward at a speed that allows the plate edges to melt inwards: if you've got it right, a beautifully smooth, rounded weld will result (Fig 3.21). Turning over should reveal even penetration (Fig 3.22).

3.20 Open-corner joints are quick and easy in thin sheet. Start by tacking so inner edges just touch all along their length and plates are at 90°, as here. Overlapping means one edge is shielded from flame heat and thus won't fuse

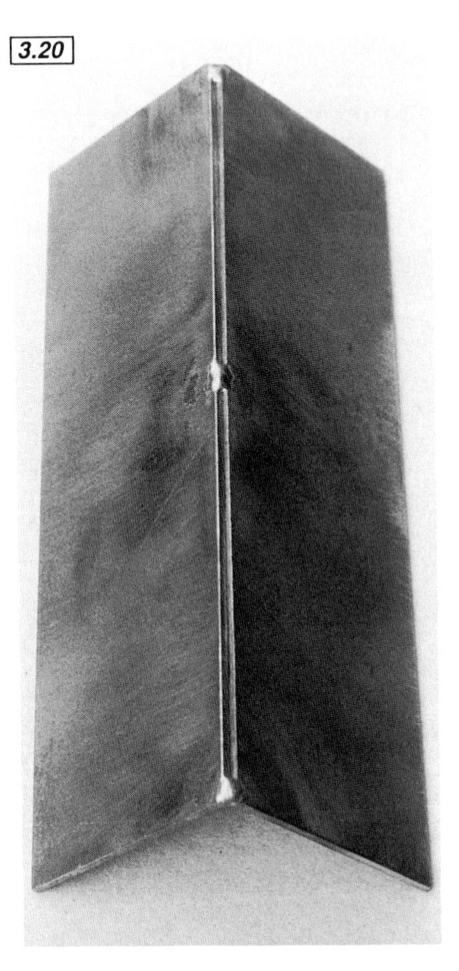

3.22 The flipside shows even penetration (A). A good open-corner joint can be hammered flat yet stay intact (B).

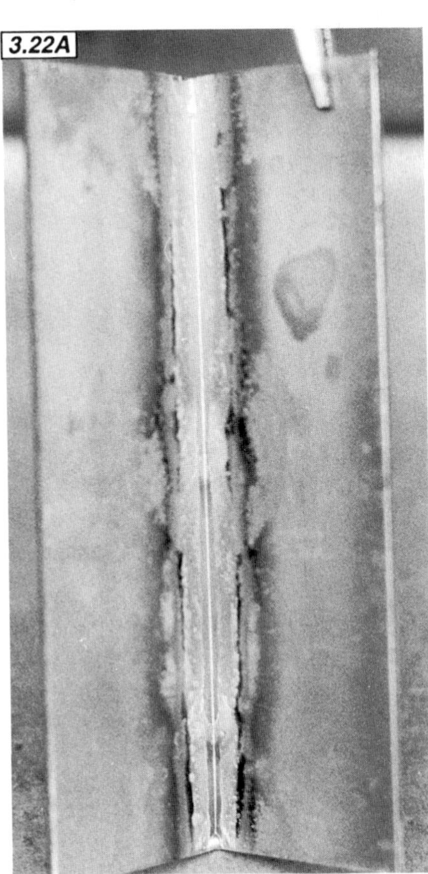

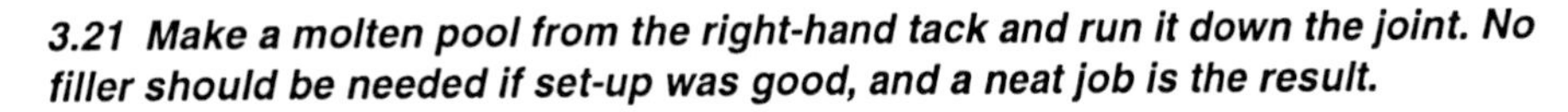
3.21 Make a molten pool from the right-hand tack and run it down the joint. No filler should be needed if set-up was good, and a neat job is the result.

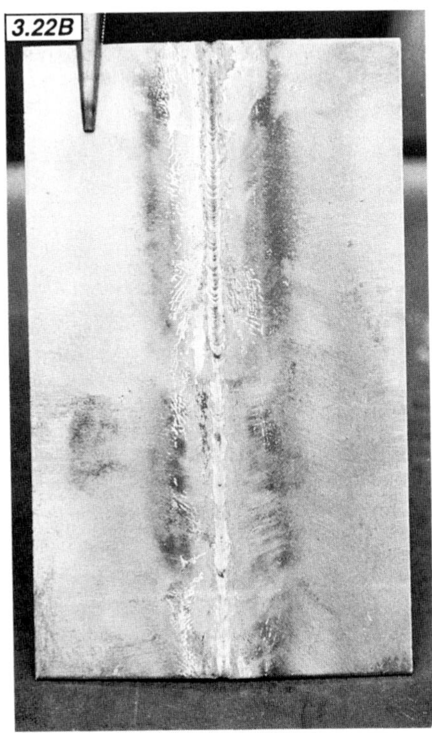

Something harder

After that easy little number, try a fillet joint. The trick here is to supply enough heat, for (thanks to the horizontal plate) the joint zone presents a bigger heat sink than a butt's simple edges.

Use a nozzle at least one size bigger than before. Ideally, support the joint so it's not resting on anything cold as this will only add to the heat sink effect.

Tack plates together on the reverse side of the joint. Bring down the flame, biasing it towards the heat-hungry bottom plate.

Start the joint by building a fillet of weld metal across the plates, making sure it fuses with the parent metal and bites down into the joint root.

Using this fillet as a starting point, move a molten pool slowly along the joint. The vertical plate is most at risk from burn-through and undercut, so be sparing with the flame up there. Adding filler rod to the top of the pool also helps cool the vertical plate locally.

The vital thing is not to rush. Be certain that fusion reaches right down into the joint root before moving forward—you'll see the pool go hollow and molten metal obliterate the root area as fusion happens. If the pool is pushed along before this is seen, all you're doing is building a buttress between the plates. The root won't be filled at all (Fig 3.23).

Having said that, sheets with good weld fusion on both faces are still usually strong enough to be hammered flat. Though incomplete fusion isn't ideal, a job only has to be strong enough for the purpose! But the job will fail sooner through vibration—or any other sort of bending fatigue—so watch out.

Getting a fillet weld right can be taxing, but is satisfying when it works. As before, work should be fast but controlled—and flame size/ manipulation is the key.

Common fillet problems come from several sources. Using too small a flame gives a tiny, fast-freezing pool that won't flow, and root fusion that's near impossible.

Too little heat is the cause of cold laps (Fig 3.14). Molten metal is just laid on top of the parent plate(s), rather than fusing in; you'll see it typically at a weld's edge. Cold shuts are the same problem, only this time happening where weld metal joins with weld metal: say, where work restarts or where tacks aren't fused into a joint during the weld run.

Burn-through or undercut on the vertical plate comes from quite the reverse—too much heat. One or both will happen if travel is too slow, if the flame spends too much time away from the joint root or if a cold under-work heatsink forces the use of an oversize nozzle in a bid for root penetration.

Lap joints are just miniature fillets. Work fast, overcoming the bottom plate's heatsink effect by using a harder flame or bigger nozzle than for the equivalent butt joint. Judge filler quantity and speed of travel so that the upper plate's top edge is just fused in to the joint.

Vertical Work

It's often possible to turn a job so it can be welded flat, which is a good move if you're not too confident

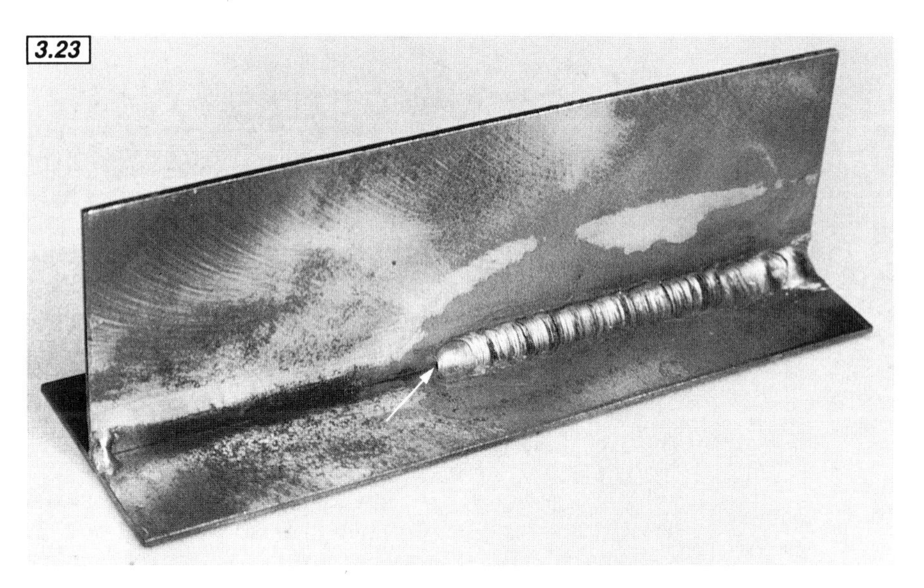

3.23 Getting enough heat into a fillet joint sees the root fused, but it's very easy to leave it unwelded (arrow)—especially if the work is rested on a cold bench. Despite such a defect, good fusion in the joint walls will mean adequate strength. Weld one and flatten it with a hammer to see, but be aware that fatigue will destroy such a joint faster than one with 100% root fusion.

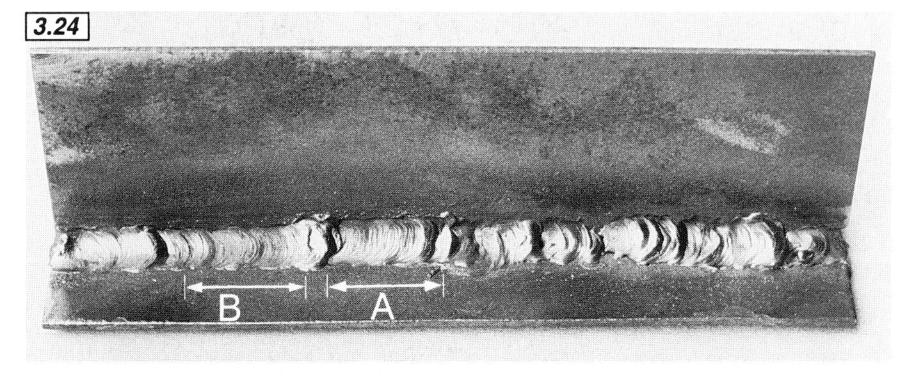

3.24 Not the best of fillets. Using too-small a flame and travelling at varying speed produced this jumble of poor fusion and misses. But things improved a bit as work progressed leftwards along the joint; small sections (A & B) look better, though undercut is on show at the top edge of (A).

about gas welding vertically. But the technique isn't that hard. Travel is generally upward, as gravity makes more of a mess of the pool coming down.

Butt joints are prepared as for flat work, leaving a penetration gap equal to or just bigger than plate thickness. Rod and flame angles stay much as they were, only shifted through 90° (Fig 3.25). The rod can either be held above the job (a hot business, as flame heat convects upward), or bent through almost 180° to form a hook and held alongside the joint.

The bottom tack is used as a starting point. Make a pool from this and add a little rod. Full-depth penetration is assured if an 'onion' or 'keyhole' is produced in the parent plates ahead of the pool (Fig 3.26). Either steepen the flame so parent metal melts and drains back into the pool, or physically push the flame down into the joint root to open up a gap.

Once the 'onion' is created, work upwards at a rate that maintains it, adding filler rod to the pool as you go. Where a penetration gap has been left in thinner sheet (say under 1.6mm) an 'onion' isn't always necessary for full-depth fusion. All this may sound difficult, but in reality it's easier to do than read about.

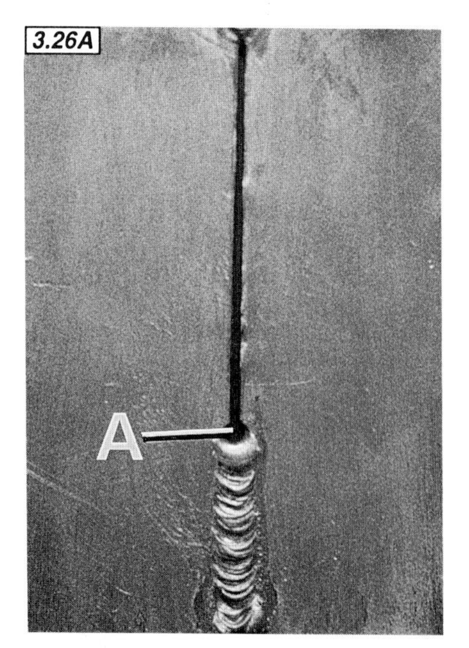

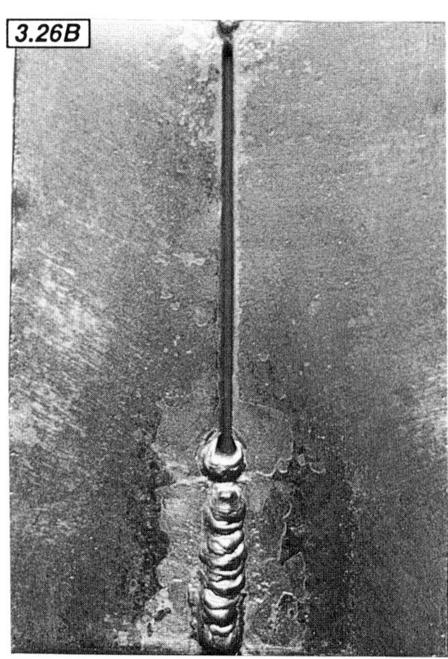

3.26 Where necessary, leave a penetration gap. To be really sure of welding full-depth, use flame heat and/or physical pressure to blow an 'onion' or 'keyhole' through the plates (A). For total penetration maintain it as the weld marches upwards (B); this is the reverse side of (A).

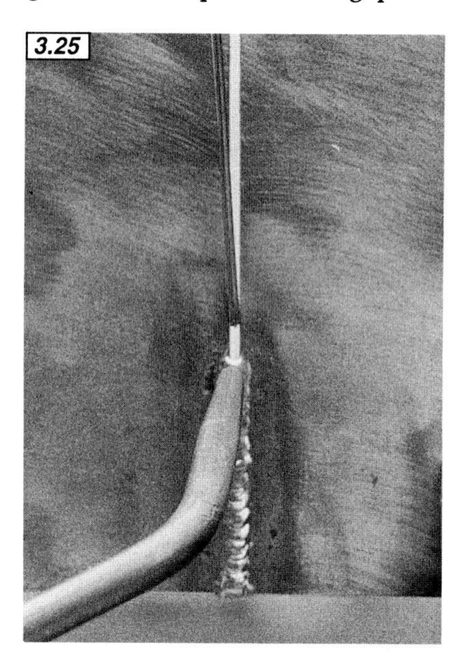

3.25 Welding vertically up with gas isn't hard. Use the same rod/flame angles as on the flat. To avoid burning fingers, either hook the filler rod and hold it from the joint side, or use it 'as it comes' held above the flame in thick gloves!

Overhead gas welding isn't so easy. There's no magnetic transfer of filler between rod and work as in MMA, and gravity wants to dump the weld pool into your lap.

This painful tendency is reduced by keeping nozzle size to the minimum consistent with good fusion, and working quickly. Rod and blowpipe angles can be steeper that normal; try bringing them towards a right-angle with the work. Where circumstances allow, turn the job to the flat.

It might even be worth dismantling something if the result is at all in doubt. The extra effort will probably counterbalance time and temper lost should a poorly welded overhead joint fail in service.

Bronze welding

A lovely pastime, this. Rather than melting parent metal, a copper/zinc alloy is used as a 'glue' to hold things together or build up worn surfaces.

Being a low-temperature process, very thin sheet can be joined—and there's less distortion than with fusion work. There are another couple of bonuses: thick things can be joined to thin things (Fig 3.29), and different metals can be welded to each other.

Five inputs are needed for a good result: *the right flame, the right heat, the right rod, the right flux and (most important) the right preparation.*

Nozzle size first. The work isn't going to be melted, only brought to red heat. So a nozzle smaller than for the equivalent fusion joint will be needed, and probably a soft flame to keep joint temperatures in check.

Flame setting is different, too. Though bronze hates oxygen, a neutral or just-oxidising flame is actually used. Why is this? Surplus oxygen in the flame encourages an oxide skin to form over the molten bronze, keeping its zinc content from volatilising off. But if too much oxygen is around, the bronze won't flow.

Filler rod type varies with metals to be joined and the job in hand. All rods are copper/zinc alloys, though exact composition is varied to change their strength and application—see Table 3.5, page 52.

General-purpose work with mild steel, copper and brass is usually accommodated by a silicon bronze rod. As brass melts at much the same temperature as the bronze filler, it's effectively fusion welded.

Where more toughness and/or wear resistance is called for, more expensive nickel bronze rods come in handy. The material work-hardens, so is good for building up worn bits as well as welding steel and malleable cast iron.

Last of the commonly farm-used rod materials is manganese bronze. Priced much as silicon bronze, this offers more tensile strength and is good for joining cast/malleable irons as well as low- and medium-carbon steels.

Flux is needed with all the above to keep slag, oxides and gases from the weld. If you're using flux-containing rods, be sure the material hasn't deteriorated; it will if rods are left to fester under a bench. Old flux crumbles or flakes away under light finger pressure.

If using pot flux, this needs to be fairly fresh for best effect; iron-hard stuff has had its day. A general-purpose bronze flux will do for all three rod types above.

Preparation is the single most important thing in bronze welding/brazing. Surfaces must be really physically and chemically clean if the filler material is to flow over and 'hook' into them, so elbow grease is called for.

Hit rusty/dirty/painted plate with an angle grinder, and follow this up with a rub-round using emery cloth. Parent plates must be shiny and squeaky-clean in and around the weld area. The final job stands or falls by the effort put in at this stage.

Flowing the bronze

Bring joint components up to dull red heat. Then weld, using much the same technique as for fusion work.

If preparation and temperature control are good, liquid bronze will spread quietly and without fuss. Experiment with heat input: see how liquid filler only moves where the parent metal is hot enough, letting close control be kept.

The chances are that first attempts will see the filler spitting and fuming. It's telling you that it's too hot; the zinc is vapourising in a huff, leaving behind weak copper-coloured residue (Fig 3.27). If overheating is less severe, weld-weakening pinholes are left by the departing zinc. A carburising flame produces the same effect.

If a weld shows pinholing, grind off the old deposit and try again. Either use a softer flame, or—if this can't be achieved without backfiring—a smaller nozzle. If there's nothing smaller in the box, control heat input by playing the flame on and off the work. Check joint edges, too. Sharp corners and projections will quickly overheat, giving local pinhole problems.

Like other forms of welding, good bronze work is a tightrope act. Aim for a rapid, controlled process that leaves finely rippled weld surface (Figs 3.27A and 3.29). If the bead is smooth and featureless, it was too hot.

Sometimes molten filler just sits around in a ball (Fig 3.28). Then the work is either too cold or too dirty; in either case, bronze won't flow. Use a little more heat and/or reclean the plates, then watch it go where you want.

Brazing uses the same fillers, flame

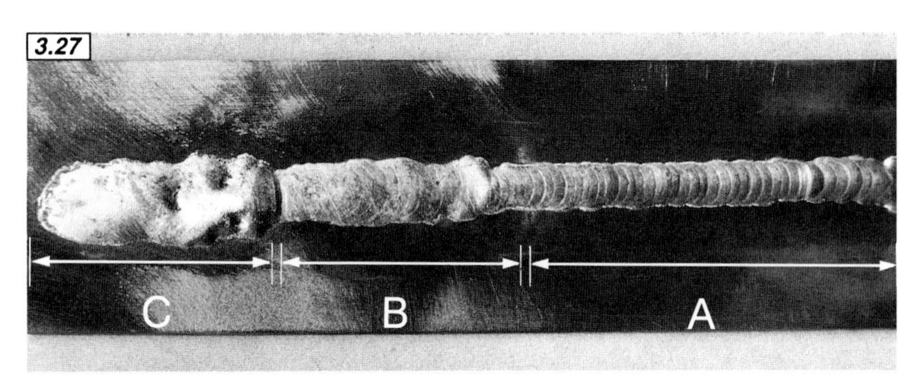

3.27 The right heat input, good fluxing and steady forward travel produces an even-width bronze bead with fine surface ripples (A)—that's what to aim for. Slight overheating pushes the bead out wider and ripples disappear (B), while overheating drives off zinc in a spitting huff, leaving a pinholed and mottled copper-coloured mess (C). Experiment with flame heat to see how hot bronze can be spread over a clean plate at will. Where the substrate is hot enough and clean enough, there it'll run.

3.28 Where work is too cool, bronze sits as a ball and won't flow. It'll act the same if the underlying plate is exceedingly dirty.

and fluxes as bronze welding. But it'll only work where a small gap exists between joint components, for only there can capillary action draw in liquid filler.

As before, excellent preparation is vital. Any dirt left behind means bronze won't flow, and the operator can't see such problems as these are inside the closed joint.

So prepare the ground by sprinkling a layer of flux between components. Some fluxes can be made into a paste with water, letting them hold on vertical surfaces.

Then bring both components up to red heat, checking that all areas are equally hot. Melt a little filler rod at the joint edge; if everything is close together, hot enough and clean enough, bronze will be quickly sucked out of sight. Carry on until the joint is full.

The nut and plate example (Fig 3.30) shows how thick and thin items can be brazed to leave a neat, strong zero build-up finish. Natural surface roughness and a preparatory flux layer provide enough gap for capillary action.

Cast irons

Bronze welding's usefulness isn't confined to steel. Castings can be repaired without bringing them up to fusion temperature, minimising post-weld cracking problems.

But the necessary clinical cleanliness can be hard to achieve with oil-impregnated surfaces. Solvents can help, but be sure all residue is gone before heating; cresylic acid-based degreasants can produce lethal phosgene gas as they get hot.

Flow a thin 'wetting' coat of manganese bronze filler on to both surfaces before joining. This 'tinning' makes sure a good casting-to-bronze bond is established. Then bring joint halves together and weld normally, using good fresh flux. In deep Vees, use multi-runs to keep down heat input.

Cool the finished job very slowly. Keep it away from draughts and cold surfaces, and it'll be all right.

Cutting and piercing

Slicing steel relies on pure oxygen's keenness to combine with preheated metal, giving out extra heat in the process.

Fuel gas can be either propane or acetylene, with propane being the cheaper but more oxygen-demanding option. Here we'll consider only oxyacetylene cutting, as welding requires acetylene fuel gas and this is likely to be on the farm.

All but sheet metal oxyacetylene nozzles have a central cutting orifice surrounded by a ring of preheater holes (Fig 3.31). They're stamped with ANM (Acetylene Nozzle Mix), followed by an imperial measurement ranging from 1/32 to 3/32 of an inch. The bigger the hole, the thicker the plate that can be cut (Table 3.6).

Sheet nozzles come in one size only, have a single heater hole and are stamped ASNM (Fig 3.37).

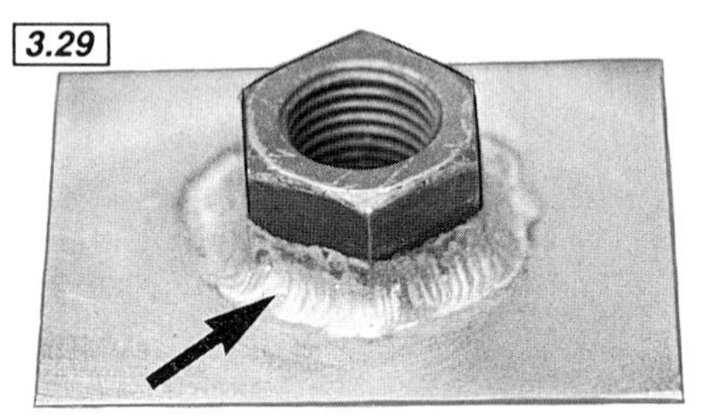

***3.29** One of bronze welding's strengths is the ability to join thick things to thin. Even, fine surface ripples on this joint (arrow) suggest temperature was held in check and rod was added steadily.*

***3.30** This now seedy-looking brazed joint has stood up to the hammer and chisel treatment. The nut was set on a thin bed of flux and brought to red heat; then the plate was heated. A little bronze was melted at each nut flat, and disappeared under the nut to emerge on the thread side. Don't spend too long getting everything hot, as oxides will build up and hinder bronze flow.*

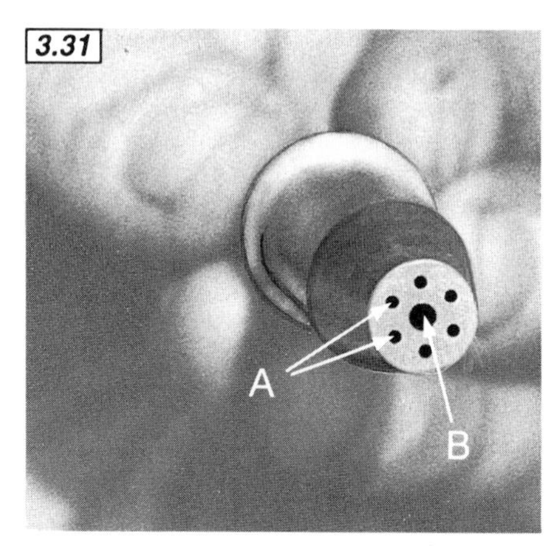

***3.31** A new gas cutting nozzle showing off its round, square-edged preheater holes (A) and the central cutting orifice (B). With time these degrade, but can be returned to pristine condition with a file and nozzle reamers. If your nozzles are kept like this and used with correct gas pressures, cutting will be easy.*

TABLE 3.6 Cutting nozzles, recommended operating pressures and thicknesses handled. (Courtesy Murex)

Stick to pressures, even if using a cutting nozzle for heating. Again, the actual job determines which nozzle is used. Note that a mid-sized nozzle in good order can cut steel 75mm thick—plenty for most needs. As a guide to gas consumption, a full oxygen cylinder at 175 bar holds about 8,500 litres, a full acetylene cylinder about 5,700 litres.

Plate t'kness		*Nozzle*	*Operating pressure Oxygen*		*Fuel*		*Gas consumption Cutting-oxygen*		*Heating-oxygen*		*Fuel*	
mm	*in*	*size*	*bar*	*lbf/in²*	*bar*	*lbf/in²*	*l/h*	*ft³/h*	*l/h*	*ft³/h*	*l/h*	*ft³/h*
6	1/4	1/32	1.8	25	0.14	2	800	28	480	15	400	14
13	1/2	3/64	2.1	30	0.21	3	1900	67	570	20	510	18
25	1	1/16	2.8	40	0.14	2	4000	140	540	19	470	17
50	2	1/16	3.2/3.5	45/50	0.14	2	4500	160	620	22	560	19
75	3	1/16	3.5/4.2	50/60	0.14	2	4800	170	680	24	620	22
100	4	5/64	3.2/4.8	45/70	0.14	2	6800	240	850	30	790	27
150	6	3/32	3.2/5.5	45/80	0.21	3	9400	510	1380	48	1250	44
200	8	1/8	4.2	60	0.28	4	14800	330	960	34	850	30
250	10	1/8	5.3	75	0.28	4	21500	760	1560	55	1420	50
300	12	1/8	6.3	90	0.28	4	25000	680	1560	55	1420	50
Sheet		Asnm	1.5	20	0.14	2	800	206	85	3	85	3

When using Type 3 cutting attachments the higher oxygen pressures should be used up to maximum cutting capacity of 150mm.

The cutting process is simple. Preheat flame(s) first bring the material up to red heat. Oxygen is then fed in via a lever on the cutting attachment: on hitting hot steel the two combine. The steel actually ignites, giving out more heat to keep the process going and turning into liquid iron oxide. This is cleared from the cut by pressure from the oxygen stream, leaving a gap or 'kerf'.

Cutting is the only job in which different oxygen and acetylene pressures are used (Table 3.6). Thus hose check valves must be in good working order!

Good cutting is almost entirely down to equipment condition and setting. Some operator skill is called for, but this is minimal in comparison with welding. Equipment condition and set-up is everything; it's fair to say that a good cut just can't be made with poorly maintained gear.

Look first at the nozzle. This needs to be as nearly new as possible—see Figs 3.31 and 3.32. The aim is nozzle orifices that are round and sharp-edged. With use they turn bell-mouthed, making the cutting stream tumble as it leaves. The result is a poor cut.

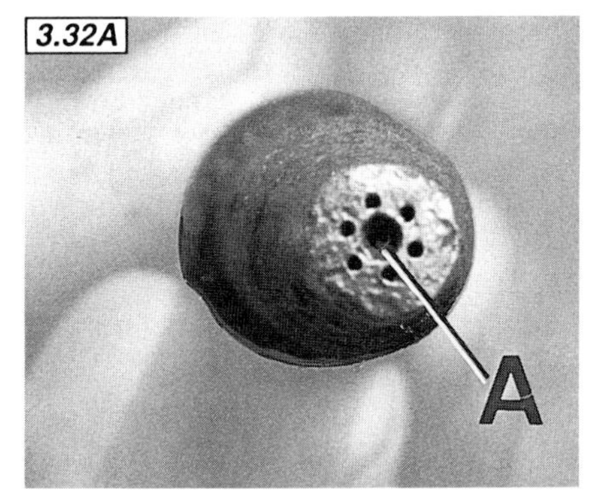

3.32 Bell-mouthing in worn cutting nozzles (A, arrow) causes turbulent flow and ruins cutting ability. Nozzle (B) has been dropped, flattening the soft copper alloy tip and deforming holes. It won't cut properly until refaced with a file and reamed.

Reclaiming worn nozzles

With both welding and cutting nozzles, the maker builds extra 'meat' into the tip to allow for refacing. Do this a little and often to keep nozzles in shape: use a smooth file and/or fine-grade emery block to dress the tip, and ream all holes gently afterwards.

To check that the job has been well done, light up and set a neutral flame (see later). The flame envelope should be symmetrical when looked at from the side and above, both with and without the cutting lever held down.

If one heater flame shows a feather or is more oxidising than its mates, ream it again and check that its internal drillings are clear. With the nozzle held up to the light and angled, you'll see through them.

Sometimes a smoky acetylene flame appears at the nozzle holding ring. It's a sign of poor internal seating and can sometimes be cured by very gently dressing the nozzle seats with worn-out emery cloth. But if seat(s) have been damaged by a fall, there's no option but to buy another nozzle. They're not expensive.

A couple of other points. It's worth keeping spare nozzles in a hardwood block so they're not sculling around and risking damage. And because they're soft, it's easy to flatten internal passages if you use a hammer to tap them loose from the cutting head. A softwood block is much kinder.

Getting set for a good cut

A balanced flame is the other half of a good cut. Set pressures according to Table 3.6, one at a time with the appropriate blowpipe needle valve(s) open. If material is really rusty, increase pressures a little. Outside work on a windy day robs heat, so nozzle choice needs to account for this.

Close all needle valves. Light acetylene first as normal, and adjust flame size so it's 'bright'. Then open the oxygen valve completely so it plays no further part in proceedings. Feed in oxygen via the cutting attachment needle valve until the preheater flames are neutral and then hold down the cutting lever.

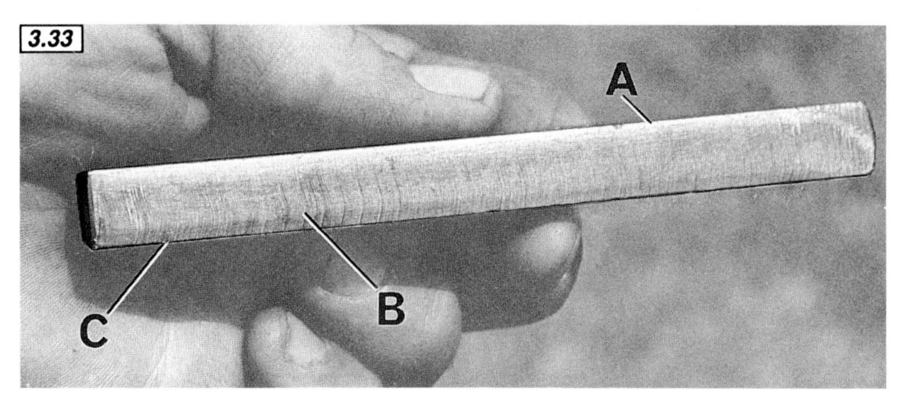

3.33 A good gas cut is one that falls on its own rather than needing to be belted off with a hammer. The clean, unmelted top edge (A) shows that preheat flames were not too big or moving too slowly. Draglines just curving at the bottom (B) suggest that the speed of travel was as fast as possible—much faster and the cut would have stopped. A clean lower edge (C) shows heat input was right for material thickness.

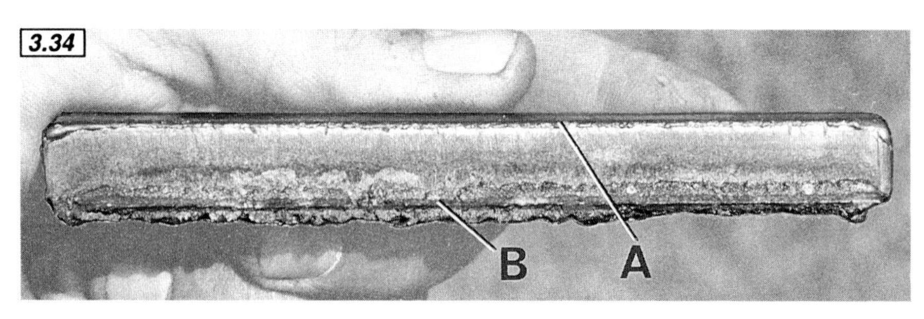

3.34 A poor cut. A melted top edge (A) and crusty slag clinging to the lower edge (B) suggest heat input was too great—probably through too big a nozzle or too hard a flame. Usually, such cuts won't part of their own accord as slag holds the kerf together.

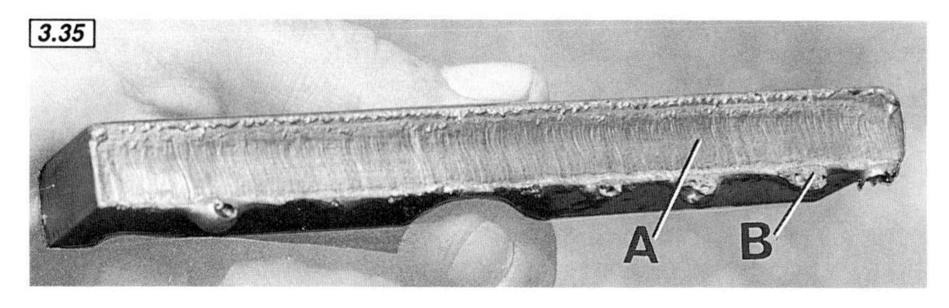

3.35 Where cutting oxygen pressure is too high, the cut face is hollowed by turbulent gas flow (A) and slag clings to the bottom (B).

3.36 Nothing much wrong here—only beading on the top edge (arrows), showing preheater flames were too low.

The chances are that the flame balance will shift slightly away from neutral. Using the cutting attachment needle valve, rebalance it. The critical thing is that the flame is neutral when cutting, i.e. with the lever held down.

A nice slice

To make a cut, bring the nozzle wholly over the start point. Keep heater flame cones around 3mm from the surface, as with a welding flame; flame heat is maximised and red heat achieved as fast as possible.

Then swing the nozzle clear of the work, feed in cutting oxygen full-tilt and quickly bring the nozzle back on target. The cut will start cleanly with minimal splashing of molten metal. With thicker sections, wait until you see molten spatter appear from under the cut line before moving off; this shows the cut has penetrated full depth.

With the cut going well, move down the intended line holding the flame cones around 3mm from the surface.

If travel is too slow the cut's top edges will melt away under preheat. If it's too fast, the cut simply stops. Should oxide splash back out at you, the cut has stopped penetrating. Either slow down and set a harder flame or use a bigger nozzle.

When a cut stops, preheat again as necessary. Where possible, restart the cut on the 'scrap' side of the job to preserve the 'best' edge.

Figs 3.33–3.36 show a good cut and a selection of problems. One very common fault is to use too big a nozzle or too hard a flame; in either case the bottom cut edge will seal with foamy, hard-to-remove oxide.

Even the smallest nozzle will cut more than you think—try it and see! Most work can be handled by 1⁄32 inch and 1⁄16 inch sizes. And as with welding, a flame can be 'softened' by reducing acetylene flow until popping starts and then dialling in a little more flow for safety. Using a softer flame is often a good ruse for curing underplate oxide build-up, especially when cutting thinner-section plate.

A clean cut shows smooth drag lines (Fig 3.33) and has little or no

3.37 ***A sheet metal nozzle has two holes—the higher for preheating and the lower for the cutting stream. Used with a guide it'll work well in material up to 3mm thick, though heat on sheet means some distortion.***

oxide on the bottom edges. What there is will chip off easily.

Tips for easier work

Before cutting heavily rusted or painted material, use flame heat and a wire brush to clean it. This makes marking-out much simpler.

Chalk will do for marking, but a soapstone stick (Fig 3.38) is much better, as, unlike chalk, it doesn't melt under flame heat. And a row of centre-punched dots make an easily seen guide for cutting circles—slice away until the inner dot halves disappear and the hole will be about the right size (Fig 3.39).

Sundry commercial attachments are on offer for cutting circles. These aside, straight-line cuts are simple if a length of angle iron is clamped alongside the line; choose a size that leaves the heater flame cones the right height above the work when the nozzle clamp nut is rested on an upturned edge. But keep an eye on the guide, as after a few cuts it'll probably warp into junk.

Start cutting a hole from the centre. Use a central punch mark as target for the heaters, for its edges come up to heat quickly. When ready, lift the nozzle well clear of the plate before gently feeding in the cutting stream, or the initial splashing will part-block the nozzle.

Working in a confined space will probably overheat the nozzle, causing backfiring. By all means plunge-cool it in a bucket of warm water, but leave oxygen flowing to stop passages filling up. Check the nozzle ring afterwards in case contraction has loosened it.

If the nozzle starts to whistle, either spatter or too-low gas flow has led to one or more preheater flames burning inside. Shut off quickly and find the cause, or the nozzle will soon melt.

Stainless steel or high-chromium welds won't cut, only burn. Cutting

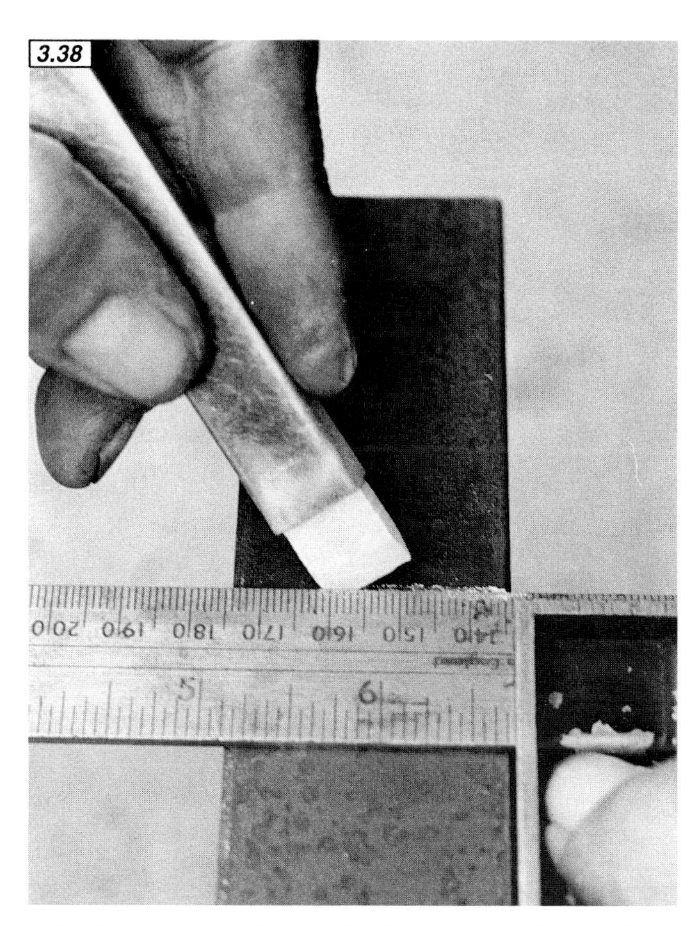

3.38 ***A soapstone stick and holder cost pennies. Cutting is more accurate as the line doesn't disappear under flame heat.***

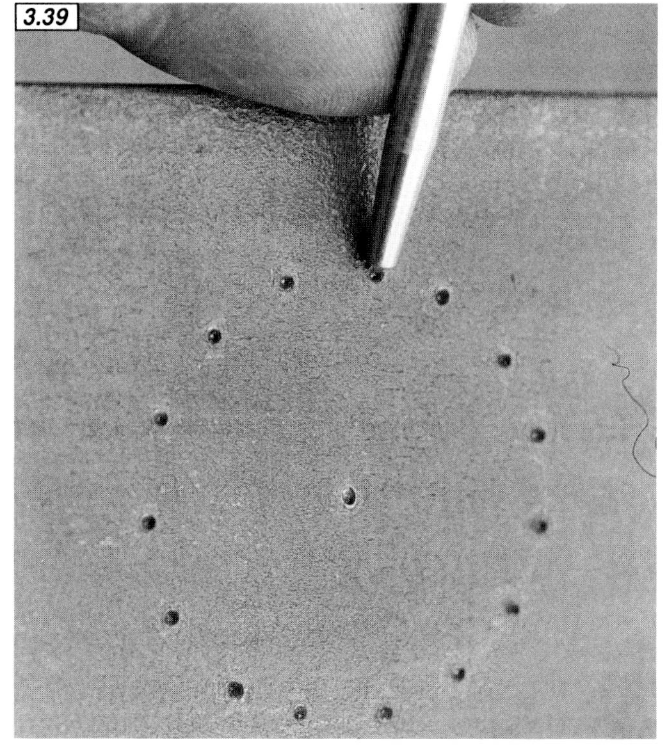

3.39 ***When cutting circles, punch-dot the centre for an easier start. Providing dots round the edge makes it easier to see during work, and taking the cut halfway through the dots brings the hole to the right size. That's if your hand is steady enough! If not, a simple pointed pivot can be clipped to the cutting head and set to the appropriate radius. Buy one or DIY.***

oxygen can't combine with such metals, making them a no-go area. Cast iron will cut, but poorly; use a carburising flame and be prepared for a hot, messy job. Let the cut penetrate full-depth before starting, weave to leave a wide clearable kerf and keep the nozzle high to minimise blocks.

Gas cuts stop very quickly if there's the smallest gap between parts. This works to advantage when slicing a bearing off its shaft or the head from a bolt, but means that stacks of plates can't be cut very well.

Section 4
SOLDERING

Soldering is essentially a lower-temperature version of bronze welding. All that really differs are the alloys used, which are less mechanically strong. They also melt at lower temperatures, and direct heat transfer from an electric- or flame-warmed soldering iron is enough to set them flowing.

The process sub-divides into soft and hard soldering. Taking the last first, hard (or silver) soldering calls for the most heat. The alloy used is a fairly expensive mix of copper, zinc and silver which, being tougher than soft solder yet still electrically conductive, is useful where a joint must stand moderate heat and vibration. It's a relatively specialist technique that won't be expanded on.

Soft soldering is a much more common lower-temperature process, used on heat-sensitive items or on joints that don't need much mechanical strength. Examples are electrical wiring, radiator header tanks and plumbing joints in copper pipe; different metals can be joined to each other, too. The alloy is a lead-tin mix, often spiced with antimony. It's flowed into or over a pre-cleaned, fluxed joint.

Just as with bronze welding, three things are necessary for soldering success: clean parent metal, the right temperature and the right flux.

Cleaning is, as ever, a matter of elbow grease. Soft solder is even less tolerant of dirt than bronze, so it's essential to be thorough. Grind surfaces where possible. Where not, polish to a bright finish with emery cloth. Dubious-looking or corroded surfaces must be scrubbed clean.

Fluxes are of two sorts, though both do the same jobs—protecting the metal from oxidation, breaking down liquid solder's surface tension to let it flow and (with acid types) giving some limited cleaning to the joint.

Paste fluxes generally have a resin base and are relatively non-corrosive. They're used primarily for electrical work and come in a tin or as a core in the solder itself.

Acid fluxes come in liquid or paste form, have a more vigorous cleaning action and are generally used on plumbing or sheet metal work. Being corrosive, they're not for delicate electrical stuff. Residue must be washed off the joint and the user's hands afterwards.

Iron options

The commonest way to supply heat for soldering is via an iron. Why it's called an iron when the working tip is made from copper isn't clear, but there it is. Generally, the bigger the copper end (or bit), the greater is the heat reservoir, so the bigger the job that can be handled.

The key thing is to match the iron to the job. If it's too small it won't supply enough heat to the work. If it's too big, access may be difficult and solder flow can be uncontrollable.

Heat-sensitive electronic components can be rapidly damaged by heat. Whatever the iron size, if there's any question of this, use long-nosed pliers as a heat sink (Fig 4.1).

The traditional soldering iron holds its copper bit in a steel/wood handle (Fig 4.2). The bit is generally heated by blowlamp or propane torch. A neutral oxyacetylene flame will do, but has to be used with great care if the bit isn't to be overheated and covered in oxide scale.

How do you know when it's hot enough? Play the flame over the bit's thickest section until the bit turns copper colour and a green flame ghosts round it. Then you're ready.

The bit's tip must be clean and 'tinned'—given a coat of solder—

***4.1** Whenever an electronic component looks as if it might be cooked, grab its leads with pliers to siphon off soldering heat.*

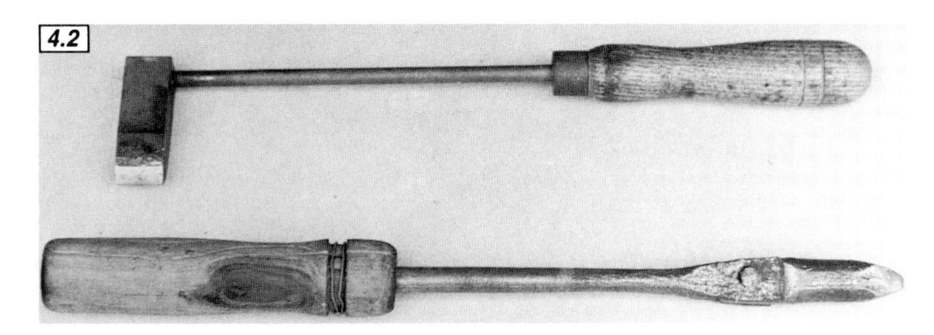

4.2 Traditional soldering irons hold different-shaped bits in a steel shaft.

4.3 No matter how the bit is heated, it must be kept clean. This dirty-looking, oxide-covered traditional flame-heated bit (A) won't transfer heat properly and brings contamination to the joint. Light cleaning can come from a wipe with a damp rag (careful!) or a scrub with a wire brush. Burnt or old bits can be reclaimed by light filing. Before use, any bit must be tinned—first fluxed, then covered with a thin coat of solder and refluxed to leave it shining silver (B).

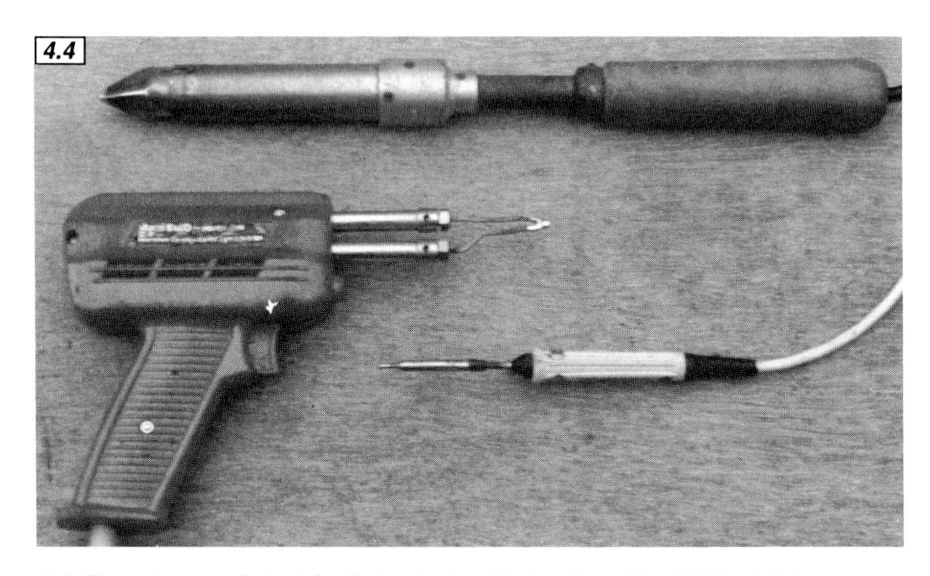

4.4 Three types of electrically heated soldering iron. The 20W miniature variety (bottom) is intended for small jobs, electronics especially. The much bigger version (top) has a jumbo-sized tip, giving a good heat reservoir for big work but taking an age to heat up. A good all-round tool is the 100W instant heat gun (centre), which quickly comes to temperature on pulling the trigger and offers interchangeable tips for different work, including plastics.

before work starts. A tatty, burnt tip won't transfer heat well and brings oxides to the work, so titivate an old one by filing back to bright copper when it's hot. Tin by sizzling in flux and then flowing a little solder over the end to make a bright silver jacket (Figs 4.3).

More common today are electrically heated irons. Small versions are perfect for wiring and electronic work (Fig 4.4). Bigger ones can be backed by fast-acting heater coils, making an 'instant heat' gun—useful to cut waiting time. At the top of the scale are jumbo electric irons, which take an age to get to full working temperature but supply plenty of heat for larger jobs.

The rules for preparing electric irons are the same. Wait until it's up to full working heat, inspect the tip for grot and clean/re-tin as necessary.

Making the connection

To show soldering's usefulness, a couple of jobs will be covered pictorially—joining electrical wires and fitting a cable nipple. It's worth remembering that like all glues, a soldered joint is at its strongest where the bond covers a large surface area and is loaded in shear rather than peel (Fig 4.6).

Electrical wiring is first. The task could be joining a broken fencer lead or replacing a tractor battery lead terminal; the principles are the same.

1. Electrical work needs a non-corrosive flux. So use flux-cored solder and back it up with a separate supply; a tin of Fry's 'Fluxite' is good (Fig 4.5).
2. Cable ends are clean enough to solder only if they're just-stripped bright copper. If not, tease out strands and rub them between emery cloth folds until they shine. Clean fingers, twist strands together and dip wire in flux pot.
3. Bring iron up to heat—a green flame for copper bits, a good smoking heat for electrical irons. Check tip for cleanliness. If okay, prepare for work by dipping in flux—watch it sizzle and smoke.
4. Lay one face of the iron—not the

4.5 ***Peripherals for soldering. Resin flux and cored solder (bottom, left/right) are mainly for electrical work. Acid flux (centre left) gives a measure of surface cleaning, and must be washed from the joint after use or it'll corrode. It's generally used with plain solder (centre right) on plumbing and sheet work. The gas blowtorch (top) has largely replaced the traditional blowlamp for heating non-electrical irons and supplying direct heat to large areas.***

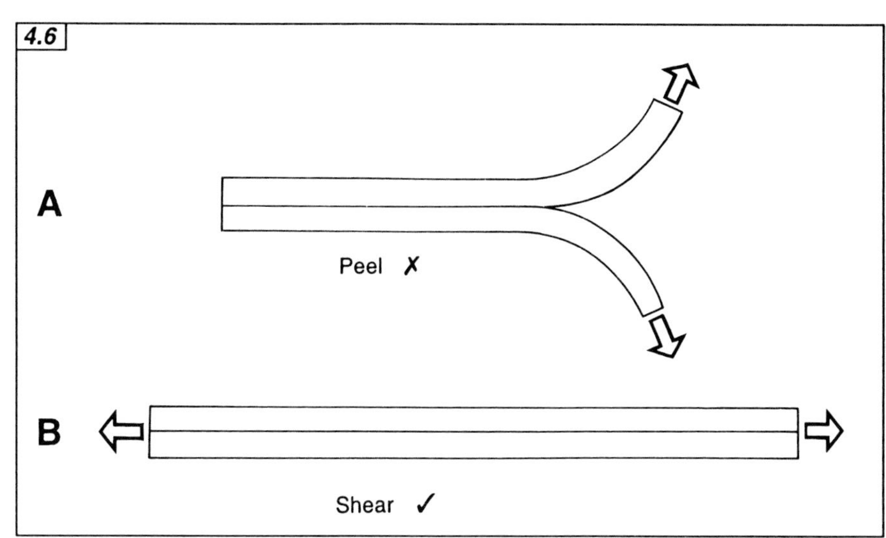

4.6 Soldering is a form of glueing using melted metal. Glued joints are weak when peeled (A)—think of pulling off a sticker. Joints loaded in shear are much stronger (B). Thus aim to design a soldered joint so it has a big surface area for good adhesion and takes only straight pulling loads.

4.7 Sizzling the bit in flux before work promotes solder flow and protects against oxidation.

point—on the cable so maximum heat flows from one to the other (Fig 4.8). The flux will melt first. When the wire is hot enough, solder will melt into it on touching. Don't transfer molten solder to the cable via the iron; oxides will form on it. Just keep the iron in place until solder melts when laid next to the bit. Patience might be needed, especially where the iron's heat-release capability is only just enough. Fill the wire with solder, checking all sides.

5. Repeat with the other joint half. Lay the two tinned wire ends alongside each other: don't twist them together, as a pull concentrates shearing force in a small area and the two will unwind under load.
6. Bring a hot, flux-dipped iron down on the joint, holding it until solder turns liquid silver and runs in both halves. If necessary (and if you've got three hands) add more solder to the joint to bulk it.
7. Take off the heat and hold wires very still until the solder suddenly freezes, turning as it does so from bright to dull silver. Any movement during solidification will give a crystalline 'dry' joint with little strength and poor conductivity. Clean off flux residue, and that's that.

Soldering a cable nipple is useful for a quick repair or when making your own controls for a one-off machine (Fig 4.9). Use the same technique for attaching 'bullet' ('snap') type electrical wiring connectors, but leave out the last two steps.

1. Non-corrosive resin-based flux is best as it won't attack cable; flux-cored solder goes with it.
2. Cut the cable end cleanly. An efficient (if rather brutal) way of doing this for tough multi-strand Bowden inner cable is to hold it between the cutting jaws of pliers or nippers, and then give these a sharp tap with a copper-faced hammer. Watch out for flying cable strands, though—wear goggles.
3. Clean and 'tin' 6–10mm of the cable end as in steps 1–4 previously. Old cable will be oil-soaked, so degrease first or solder won't take.
4. Slide the nipple down the cable so 3–4mm is left sticking out; setting the cable in a vice makes this easy. Hold a hot, fluxed iron on the nipple and cable until both are smoking and fresh solder flows down between them—patience is needed again. Continue until solder appears at the nipple base and its upper hollow crown is

4.8A

4.8B

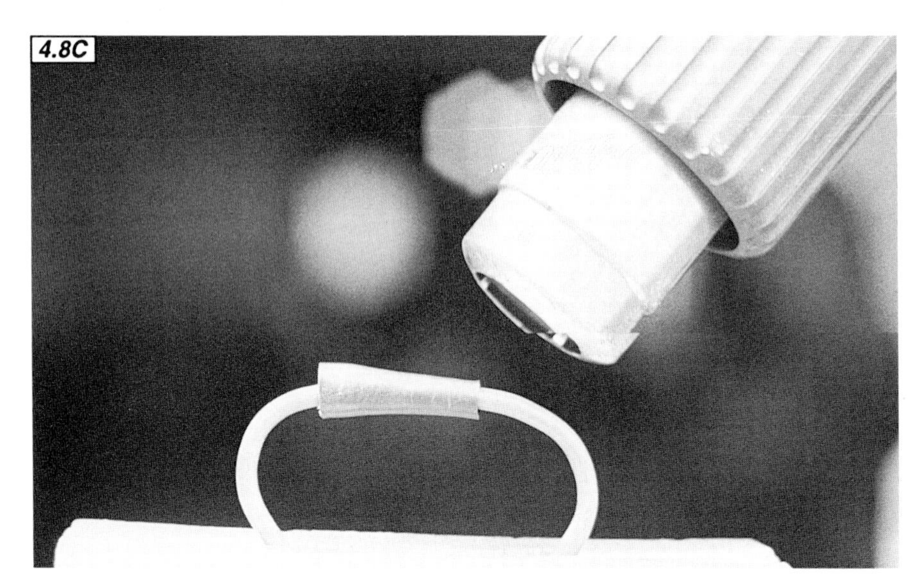
4.8C

4.8 Steps in soldering a joint in electrical cable. Clean, stripped wire is first dipped in flux and tinned (A). Borrowing a spare arm from a nearby octopus, the ends are laid alongside each other, brought back to solder-melting heat with the iron and a little extra alloy added to the joint (B). The end result takes a fair pull; twisted wires tend to unwind under this load. To insulate the joint, slide on a heat-shrink tube—available from Lucas—before soldering (C). Then heat it by match or, ideally, hot air gun to clamp it down over the bare area. Much neater than a mess of insulating tape!

4.9 Soldering a cable nipple starts with tinning a clean-cut cable end. Take tinning to below nipple depth (A). Slide the nipple over the tinned end, resting it on the vice jaws so 2–3mm of cable sticks above the hollow top. Sharing iron heat between the cable and nipple, bring both up to a smoking heat. Add a little solder (B). With a knife, splay out the cable strands to the width of the nipple cup (C). Take out the cable, turn it over and heat the cable and nipple base, flowing it in a little extra solder if necessary to give a good bond (D). Set the cable back in the vice, bring it up to heat and melt solder into the cup until it's full, entombing the splayed strands (E). Finish off by grinding back any projecting strand ends. If you've forgotten to slide the cable into its outer sheath or cut it to the wrong length, kick yourself and try again.

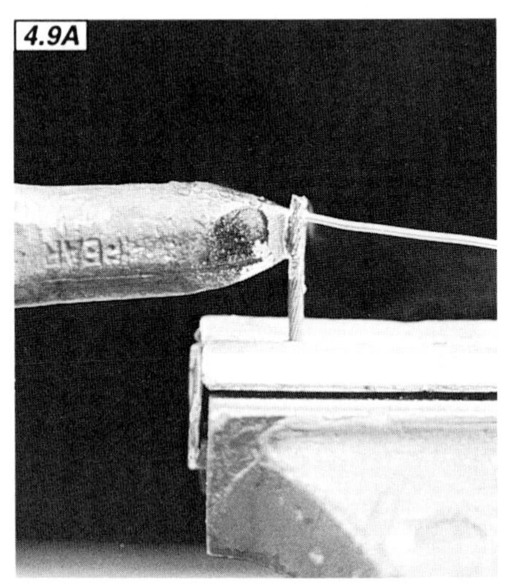
4.9A

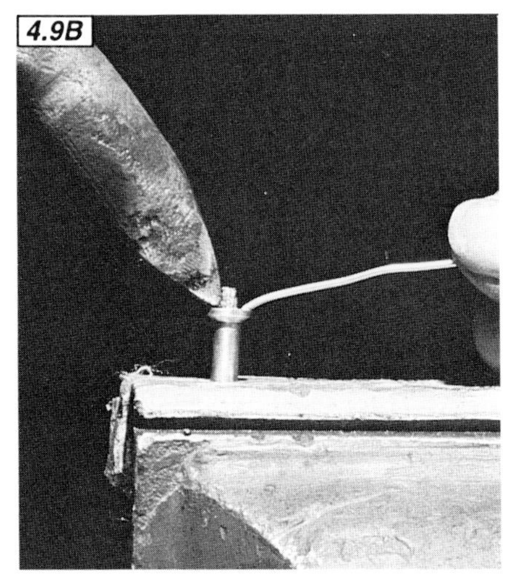
4.9B

4.9C

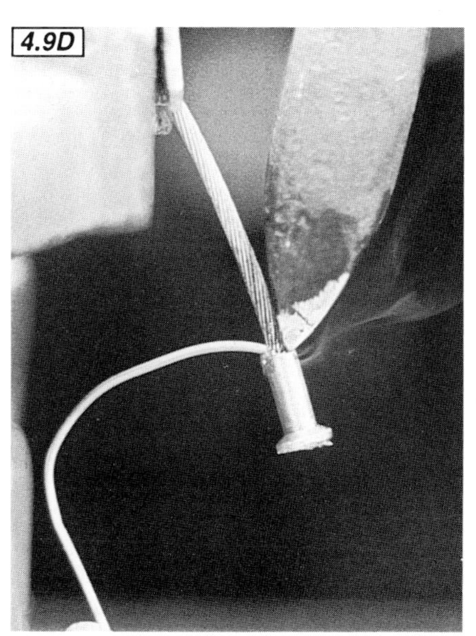
4.9D

4.9E

part-filled.

5. Working with a knife blade on the protruding cable tip, separate out individual wire strands and bend them back. This is critical for strength: the wire 'tree' thus formed won't pull out easily once it's entombed in solder.
6. Reheat the cable/nipple until the existing solder melts; then flow more into the nipple cup until the cable ends are buried. Take off heat and hold steady until solder sets.

Soldering copper pipe fittings is a useful skill.

1. Acid flux's vigorous cleaning action is valuable here. 'Yorkshire' fittings already have solder contained in a raised ring by every outlet. If more is needed (and it shouldn't be), use plumber's solder in stick form, making sure this is not covered by an oxide film. (Emery cloth will get it off.)
2. Instructions on the flux container usually say no preparation is necessary, and it often isn't. But to save the possible trouble of redoing the joint, a scrub round the pipe end with emery cloth never comes amiss.
3. Wipe a little flux round the pipe end and slide the fitting on until it 'bottoms'. Using a blowlamp or propane torch, heat the fitting evenly round the raised area where the solder lives. A non-asbestos backing pad (available from plumber's merchants) protects nearby plaster and paintwork if an in situ repair is being made. When everything is hot enough, solder will melt and be sucked along the joint by capillary action. Watch for it appearing as a bright ring where pipe and fitting meet.
4. Check that solder is visible all round the joint. If not, heat the deficient area gently; solder will usually appear. If not, reflux, heat and add a little solder from the stick. If the work is hot and clean it'll be sucked into the joint. Take away the heat and hold the joint steady until solder dulls.

Common soldering problems

These always come down to 1) a lack of heat, 2) contamination and/or 3) a lack of flux.

- **Symptom:** Solder won't melt on to the joint or melting stays local, next to the tip. Solder flows sluggishly or not at all. **Cause:** Insufficient heat. **Cure:** Check tip is clean, tinned and not covered in heat-transfer reducing oxides. Use flat face of iron for maximum contact area. Don't be impatient; heat transfer takes time. If none of the above work, the iron is either too cool or too small for the job, and the end result will be a weak joint.
- **Symptom:** Joint is hot but solder won't flow over it and may leave uncovered 'islands'. **Cause:** Surface contamination or inadequate fluxing. **Cure:** Stop trying and clean target until it shines. Coat with flux and try again with clean, hot, fluxed iron. Poor solder flow can also stem from using dull grey, oxide-covered material. Shine the solder's surface with emery cloth and try again.

Section 5

SIMPLE BLACKSMITHING

A time was when the twin strengths of a blacksmith kept a farm going: his farrier's skills kept horses at work, and his forging ability shaped and repaired iron. All manner of things passed across his anvil; bigger farms had their own smith, while smaller outfits employed the village man.

But time passed. Tyres replaced hooves and mass-production methods undercut handwork; the electric arc set turned every man into his own welder, and forges grew cold.

Too useful to throw away, many anvils ended up in farm workshops. The sound of a few blows (repeated at strategic intervals) has always been a good way of convincing a passing boss that great works are underway.

Along with the tools of the trade, practical skills also remain. Frank Dean has been a blacksmith all his life, and his sons now work in the Sussex smithy established by their

5.1 Air blast for the fire comes from the central tuyère or tue. First job when lighting the forge is to rake out any fused clinker produced by the intense heat at the fire's heart (arrow); then excavate a pit to below tue depth. Check that tue's exit hole is clear, and set a gentle draught on the blower to keep it so.

Light the fire. Pile on fuel (ideally smithy 'breeze', but if your coal merchant has never heard of it/can't get it, Sunbright singles will do), forming a central banked mound. Lay extra around the circumference for raking-in later. Don't let the fire burn hollow: keep it banked up on three sides, leaving clear space on the working side. If the fire spreads out too far during work, contain it by damping-down the edges.

5.1

5.2 The tue in big forges is usually water-cooled and relies on thermal movement for circulation. Check that the reservoir is topped up. A simple forge is not hard to make: an old water tank or big lorry brake drum supported on an angle framework is a good starting point. Its tue need not be cooled—50mm bar, counterbored to give an air passage, will do. Draught can come from an old vacuum cleaner motor with a simple flap valve to control blast.

5.2

grandfather at the turn of the century. So tradition continues. Who better to pass on some simple techniques for those with hammer and anvil and time to learn?

The written word always comes a poor second to direct tuition from a skilled man. The pictures here give guidance: best to try the techniques for yourself, using the pictures as a target.

Equipment and basics

If you're lucky enough to have a forge, then you'll appreciate the ease with which items small or large can be brought to red heat. If not, all of the jobs shown here can be managed with the gas torch and a modicum of patience. It just takes longer and costs more in fuel.

If gas is to be used, choose a cutting or heating nozzle that puts heat quickly into the job. A hearth made up from old firebricks helps to hold warmth round the work.

Getting comfortable is important. Arrange the anvil so its table is directly under your working hand: the bick (Fig 5.3) is then to the left of a right-handed man, and vice versa.

Strike blows from the elbow, rather than the wrist. Along with a proper grip (Fig 5.6), this promotes effective control of the hammer head. Work in a steady rhythm— frantic activity only makes you tired—and don't let the work get too far out of shape; correct large misalignments at the end of each heat.

Light taps on the anvil between strings of blows serve to keep a rhythm going. But when using punches, avoid a preliminary 'sizing up' tap: this usually only serves to jog the point off-target.

Finally, keep the anvil clean and free from scale and the floor clear of tools and scrap. Tripping up with hot metal is at the very least painful.

Taking heats

Recognising when steel is at a good, easy-working temperature is fundamental. The best way to discover this is by experiment: try working a piece of bar, and see how its plasticity increases as surface colour changes. It's all hard work at dull red; blood red is better, and cherry red easier still. By near-white, the steel works easily.

But take it one step too far, and the work becomes incandescent with sparklers flying off in all directions. A few degrees more heat (or seconds in the fire) and the metal crumbles away to nothing.

The plan

Decide a working strategy in advance. Think which part of the job needs to be tackled first: it's often difficult (or impossible) to do one job after another is completed. For example, taper a bar before forging the taper into an eye—the reverse is impractical. Also consider which part of the anvil to use, and which way to present the work to it. Puzzling this out on the spot loses valuable working heat.

In Figs 5.1–5.6 we start with 'naming of parts'—a look at the forge, anvil and associated tools. Then the first job is done—drawing a round bar down into a point. This is a simple job in its own right and a useful jumping-off point for all sorts of other operations.

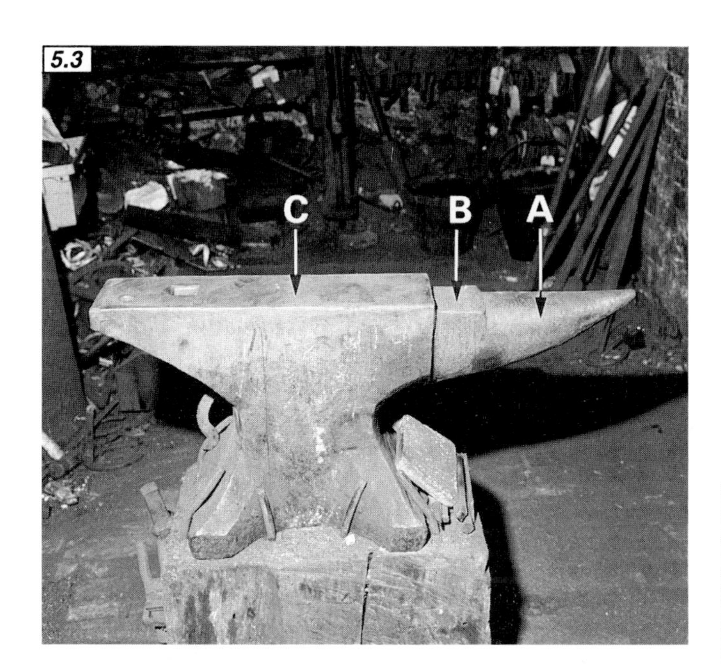

5.3 This 2½cwt London-pattern anvil sits on an elm post, and there's as much of that under the floor as above it. Working from right to left, the anvil has a tapered bick (or beak, or beck: depends where you come from—A). Then comes the step (B): this is left soft during manufacture, to provide an area where a cutting-off chisel can be used without damaging itself or the anvil. The main flat working surface is the table (C), which contains two hardie holes: the round one primarily provides a clearway for punching operations, while the square one locates matching tools securely. A farrier's anvil has no step.

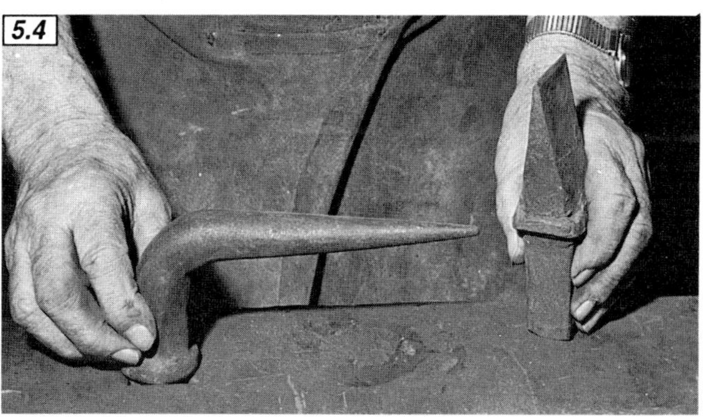

5.4 Commonly used tools for the hardie hole. A chisel-faced hardie (right) is used under hot metal when cutting, so that metal is sliced between it and the hammer face. An anvil stake provides a small-scale bick for finer work. A twin-prong 'horn' tool can be used for bending material.

5.5 Tools that you can't manage without—a selection of tongs. These must fit the metal to be held, or it'll slip about; that's annoying and possibly dangerous. This small selection was culled from Frank Dean's rack. From left to right, they are: Bolt tongs with a flared section for the bolt head or any shouldered work. Goose-billed tongs for holding flat rings and the like. Small close-bit tongs for horseshoes or jobs where there's no great length of material to catch hold of. Two pairs of hollow-bit tongs; grooved to accept round bar, the example on the right has one three-sided jaw to better accommodate square material.

Tongs can be adapted to fit different-sized material by heating and bending their reins (handles). In work, always keep tongs cool. Secondhand tongs sometimes turn up at farm sales or can be bought new. Experienced smiths make their own.

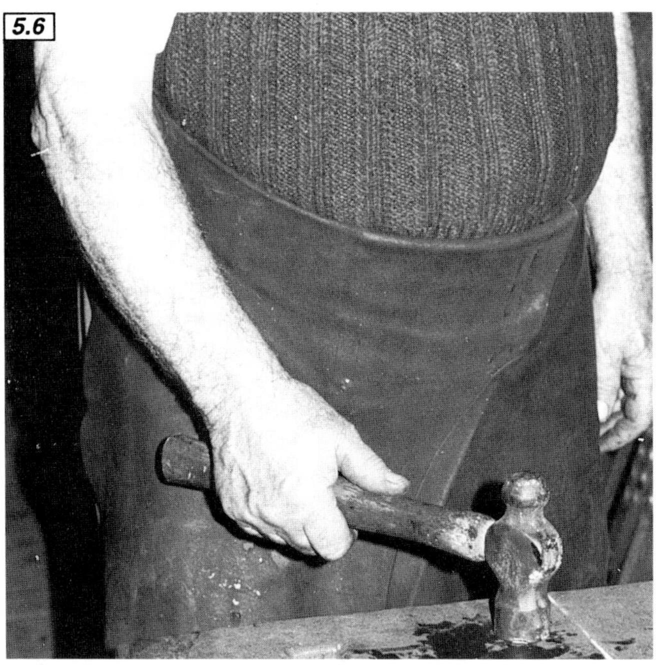

5.6 Getting a grip. Frank's hammer is nicely balanced in his hand: his thumb is on the top side of the shaft, pointing towards the head. This allows good control and permits rapid rotation when the pein end is to be used. "Holding the hammer too close to its head means you have no control", says Frank, "and working next to hot iron means you burn your fingers! Holding it at the far end of the shaft is as bad. Choose a hammer shaft that's nicely oval, and lies comfortably in your hand. Keeping the hammer balanced means there's no need to grip it tightly." Most of the hammering action should come from the elbow, not the wrist.

5.7 Round, tapered points or chisel ends are easily made by the process of drawing-down—a traditional way of giving new life to harrow tines. The metal is brought up to heat, and held at an angle to the anvil: the steeper the angle, the steeper the point will be.

See how the metal is held near the edge of the table, and the hammer face is angled to form the taper. The metal is thus shaped on two sides simultaneously. Between blows, the work is rotated—always in the same direction (in the manner of a lathe) to draw the metal down evenly. Big-section material is worked first into a square, then an octagon, then a cylinder; smaller metal can be tapered directly.

To produce a long, gentle taper, first draw out the tip area and then use the bick to speed the long work. Its rounded profile draws out metal much more quickly than the flat table. "Worth a second hammer," Frank Dean says. If the work's tip splits, take a very good heat and draw it back together with light blows. With a little practice, multi-sided material can quickly be made round or vice versa, using just the hammer and table.

5.8 The finished article. A coarse wire brush used on the hot work shifts scale, leaving a good finish.

Making eyes

A few general points to consider. Where appropriate, it's best to begin operations with straight material. Then attention can be given solely to the area being worked, rather than being distracted by problems that were in the metal before the job began. So 'true' the material up first, and if things start to go too far out of shape during the job, stop and put them right.

When working, consider where best the hammer blows should fall. If the intention is to elongate metal, then make a sandwich of the work between the hammer and anvil—this displaces material to change its section.

But if the intention is to bend material, then position the work and hit in such a way that the blow has a bending, rather than crushing, effect. In other words, hit 'off' the anvil. Figs 5.11 and 5.12 should make this clear.

Finally, two points on marking out. To know how much metal to allow when making a ring of a given size, multiply its desired internal diameter by 3.14 and add the thickness of the metal. If you're working to a desired external size, subtract the material thickness from this and then multiply by 3.14.

If precision is not vital, the anvil table can be used as a rough guide when marking off lengths. It's generally 125mm across, allowing good guesstimates to be made.

A mark often has to be made so you know where an operation (like bending) is to begin. Centre-punch the spot so that the mark falls on the inside of a bend. That way, the punch mark is closed up as the metal goes over. If made on the outside, the dot will elongate into a nasty knick point and leave a weak spot for failure.

5.9 Step one in making an eye is to true-up the metal, in this case 12mm bar. Notice how Frank is using the hardie hole as a focus for straightening, and lifting his left (or 'back') hand a little to give more clearance. Both actions provide the metal with somewhere to go. "It's hard to straighten anything satisfactorily on a flat surface", he says. "Hitting down onto a good anvil's table doesn't usually achieve much, though it might if it's worn to a saddle shape."

5.10 Unlike making rings (which are usually of a specified size), eyes like this seldom need to be made to precise dimensions. Sizing is adjusted so the eye looks right in proportion to the bar's thickness and length. Estimate a suitable internal diameter and mark the bar according to the calculation given in the left-hand column on this page.

Then shaping begins. The first 125mm of the anvil's table beyond the step generally has rounded shoulders, so work in this area to avoid putting a nick in the underside of the bar. With the metal heated to bright cherry red, use the hammer's pein to strike the first blows as shown. Work just off the table's edge so the bar's free end kicks upward, and the bend starts to form automatically. But don't work too far off the table, or a straight portion is left just beyond the start point. Aim for a gentle 'S'-shape.

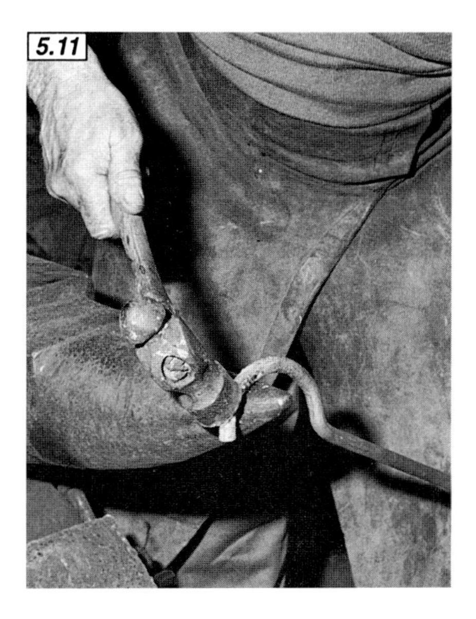

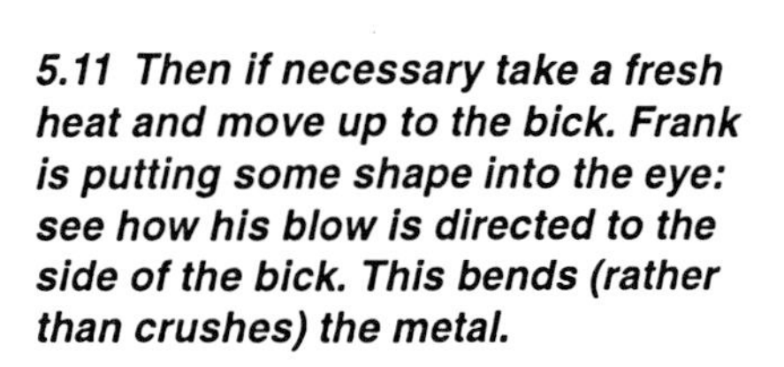

5.11 Then if necessary take a fresh heat and move up to the bick. Frank is putting some shape into the eye: see how his blow is directed to the side of the bick. This bends (rather than crushes) the metal.

5.12 Next, roll over the bar's tip. It's impossible to do this once the eye starts to be closed down. Again, the hammer is striking off the bick, rather than straight down on it: Frank's back hand is tipping the bar to provide clearance.

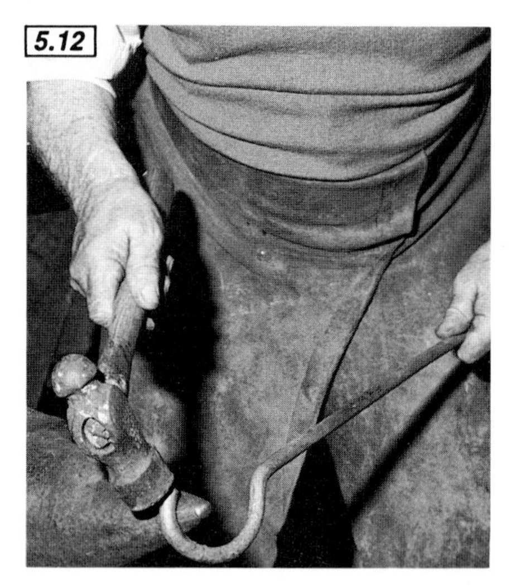

5.13 With the tip turned over, it's time to close the eye. Start by bringing it round the bick.

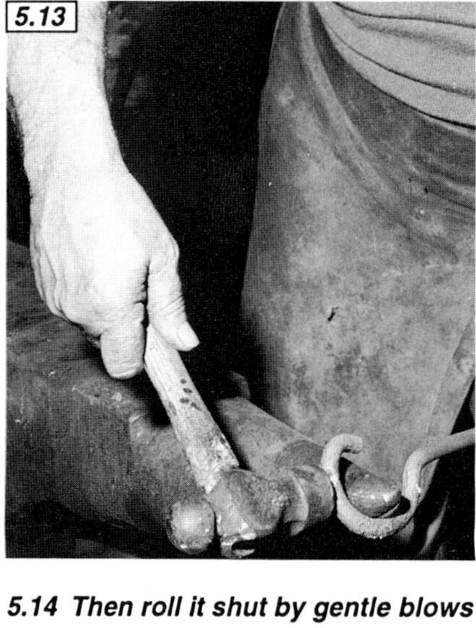

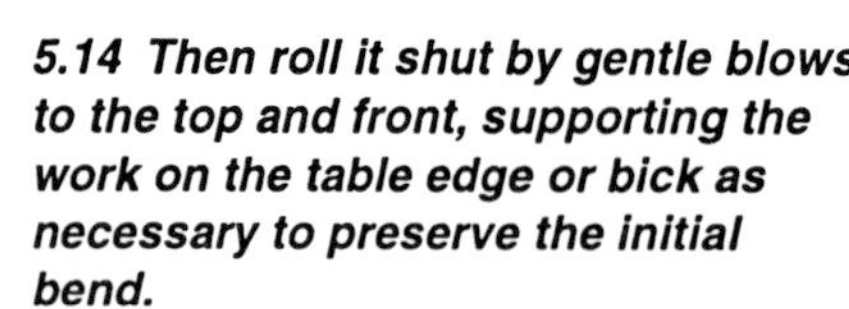

5.14 Then roll it shut by gentle blows to the top and front, supporting the work on the table edge or bick as necessary to preserve the initial bend.

5.14B

5.15 The finished article has a pleasing symmetry, but this beauty is more than skin deep. If the eye is to carry a load, this must fall in direct line with the bar's length or the eye will tend to open. So the eye has been centred over its supporting shaft—the same goes for hooks.

Edge bends

Persuading a flat strip to bend against its width is not hard if you go about it the right way (Figs 5.16–21). The process shown here produces bends with a radiused outer corner, as metal on the outside of a bend has been stretched. If a square corner is required, it is necessary first to thicken the metal locally at the bend point by 'upsetting' or 'jumping-up'. This is covered on page 78.

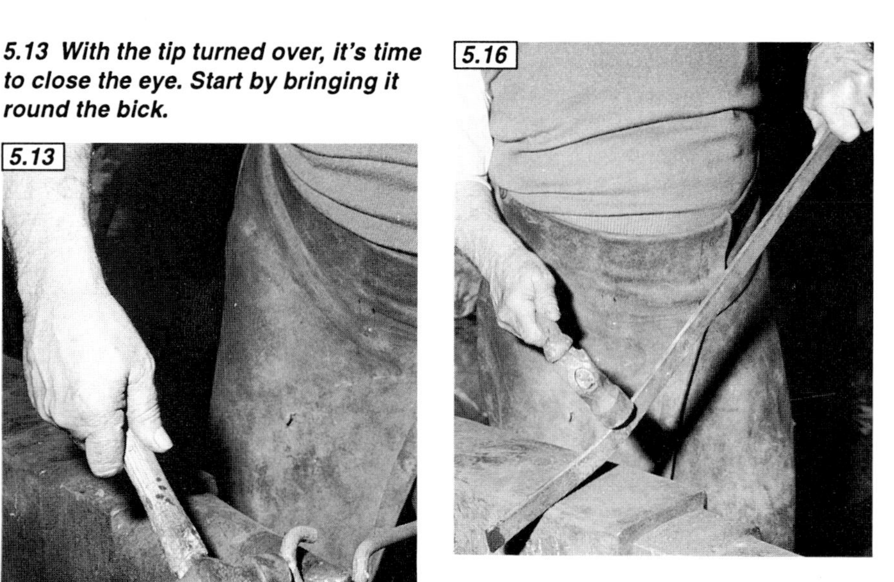

5.16 Bring the bend area to heat, localising this by either pushing the work through the fire so its tip rests in the cooler zones beyond the core, or quenching the end before work starts. Gas users have no such problems! Begin forging by putting in a set as shown: if the work bows sideways, bring it back to flat with gentle taps.

5.17 Then form a rough bend over the bick. Once more, note how Frank's hammer will fall on the work's free end: the bick is just a fulcrum for the bend.

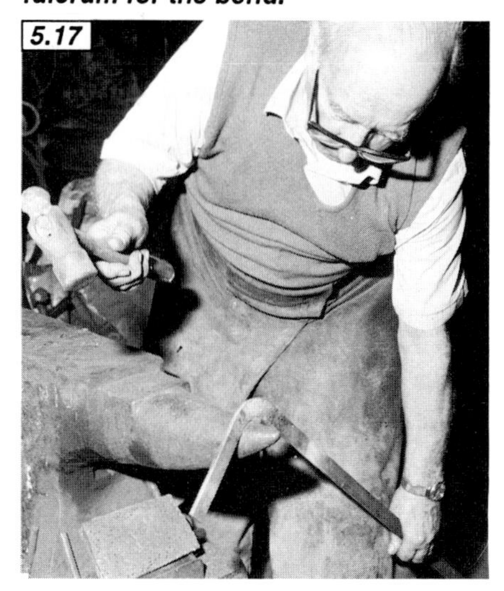

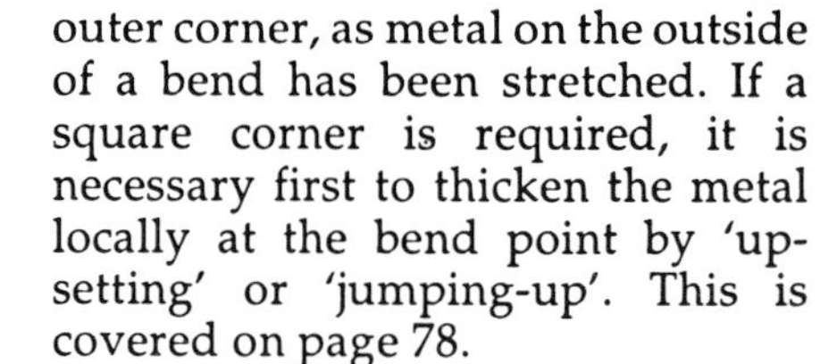

***5.18** At this stage, it's most tempting to lay the work on the anvil and try to form it over the corner of the table. But don't—repeated blows as shown will only draw the metal thinner, necking it down. You want it to bend, not elongate.*

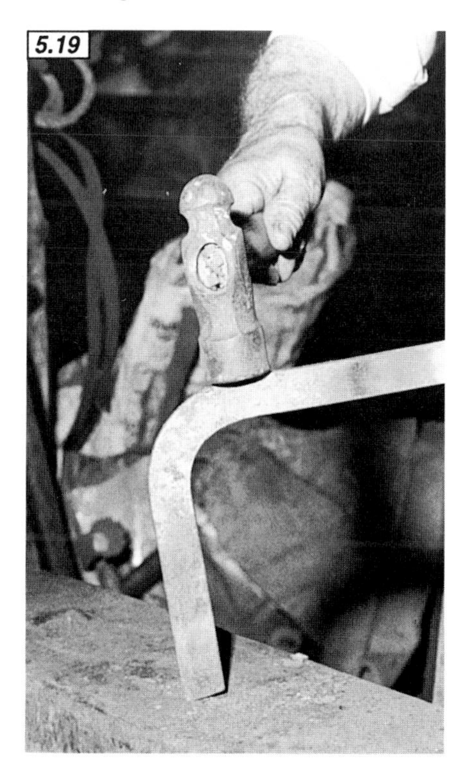

***5.19** The rough-formed bend has a large radius. To make this tighter, i.e. move towards a squarer corner, metal needs to be pushed up towards the corner. To do this, first imagine a diagonal line passing through the corners of the desired bend. Then, resting the work on the table as shown, strike just off that line. Frank's blows will tighten the bend by pushing the top leg down.*

***5.20** Work alternately on the top and front legs of the bend, driving them together and so pushing the joint closer to a right angle. But watch it: too much effort will pucker the inside radius into a 'gall', so work gently from both sides. The bend will probably want to bow sideways, so correct this on the table before it goes too far.*

***5.21** As the job progresses, out comes the try-square (A). The initial bend was too acute, so a couple of light taps were needed to bring it true. See how the anvil is again used as a fixed rest while the hammer does the work (B).*

Developing a feel for where best to position the job and strike the blow comes only with practice, and an experienced smith makes it look so easy. It's a bit like the man who charged £100 for five minutes' work in knocking out a dent in a boiler. "£5 pays for my time," he explained. "The rest is for knowing where to hit it."

Punching and upsetting

Before twist drills and power tools were commonly available, the blacksmith produced holes by a combination of steel drift and muscle power. It's still a useful way to make big holes . . . or big holes bigger.

Square holes, rectangular holes, round holes—the old technique may be slower than the drill, but it's more flexible. Neither is it wasteful; a surprisingly small chip of material is ejected as a hole forms. Instead, metal is displaced by the punch to leave a void that mirrors punch shape and dimension.

Tools for the job are readily made (Fig 5.22). A selection will accumulate with time, allowing a range of holes to be produced. Punch shank length is a compromise—too long and it can bend, but too short and the hand that holds it will be singed.

Slot punches are the first line of attack. These do the hardest work, so need to be made from the best material—something combining strength with resilience.

Frank Dean suggests that an old worn file makes a good starting point. "Take out brittleness by bringing it up to a uniform blood heat, forge out any serrations, and then

5.22 A selection of punches, which will be used in order of appearance. Right to left: slot punch, flat-ended mandril or drift, and two different-sized mandrils. All these are easily made and ground to suit different jobs. A smith accumulates a wide selection.

5.23 Frank Dean here demonstrates punching technique by putting an eye in the end of flat stock, though the method applies equally to slots and round holes partway along the length of a job. In the latter case, mark the target area by a row of punch marks. These will remain visible when the work is up to heat.

First operation with an end-eye is to round-up the corners, working the metal between hammer and anvil as shown.

5.24 With the corners rounded and a friend to hold the work, punching begins. Frank uses the anvil table, striking hard to drive the slot punch quickly. After three or four blows, the punch is cooled out.

Initial punch positioning is critical for final hole location—the punch must be centred accurately if metal is to be displaced equally to either side.

5.25 In goes a pinch of coke dust, taken from a pile on the anvil step. See how much heat is in the work: it's just short of sparkling white—ease of punching goes up with temperature. On thick work, the punch must be cooled out after every few blows.

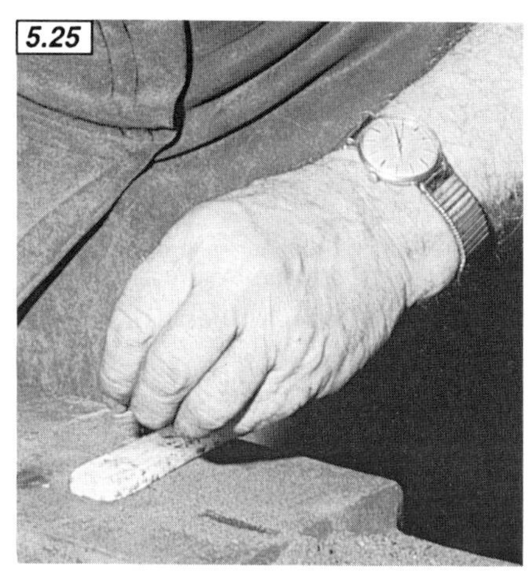

5.26 Frank drives on until hammer 'feel' tells him that the table is nigh. Then the work is turned over, and—using the coke dust mark as a guide—the slot punch is driven through again. Here he's working over the table hole to provide the punch with clearance.

cool it very slowly under a bed of ashes or lime to prevent air-hardening," he suggests. "You can then forge and grind the tip to the size needed."

It follows that slot punches must not be allowed to overheat in work. In thicker material, heat must be taken from the tip after every few blows by dipping it quickly the water bosh (trough) after every few blows, but not long enough to quench the punch back to cold.

Sticking will be a problem in thicker material. A pinch of coke dust in the embryo hole will help the punch let go, and the black dust makes a readily visible marker through thin, hot metal when the work is turned over to finish operations.

If it's round holes you need, one or more mandrils (or drifts) will follow the slot punch. These enlarge rather than pierce, so can be made from mild steel of the desired size. First to be used is a flat-ended mandril, which takes the slot gently to a circle. Any subsequent enlargement is by tapered drift.

5.27 This tiny sliver of metal is all that's lost from the slot—most has been pushed sideways.

5.28 Now follows the flat-to-round mandril, again used over the anvil hole. This tool gently takes the oblong slot to a circle. Bigger, tapered mandrils can follow until the hole is of the desired size. Surprisingly good fits can be obtained this way.

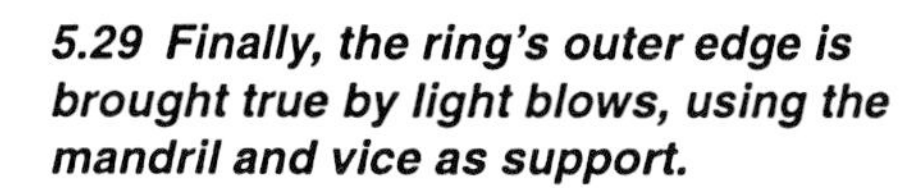

5.29 Finally, the ring's outer edge is brought true by light blows, using the mandril and vice as support.

5.30 There's a twist to this ending: with the eye completed, Frank is ready to put a 90° set below it.

5.31 Heat must be localised if the work is to bend only where required. First the whole end is brought to heat in the forge and then the target area is isolated by ladling cold water to either side. This technique is used often to confine metal deformation to where it's required.

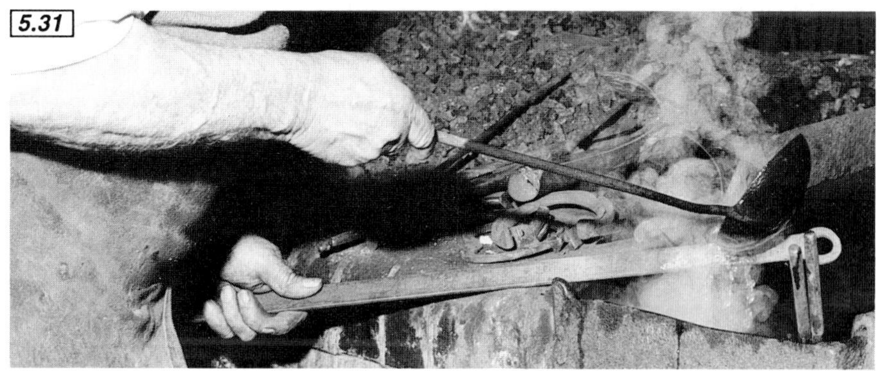

5.32 Into the vice, and a quick twist with the tongs sets a right-angle in the shank . . .

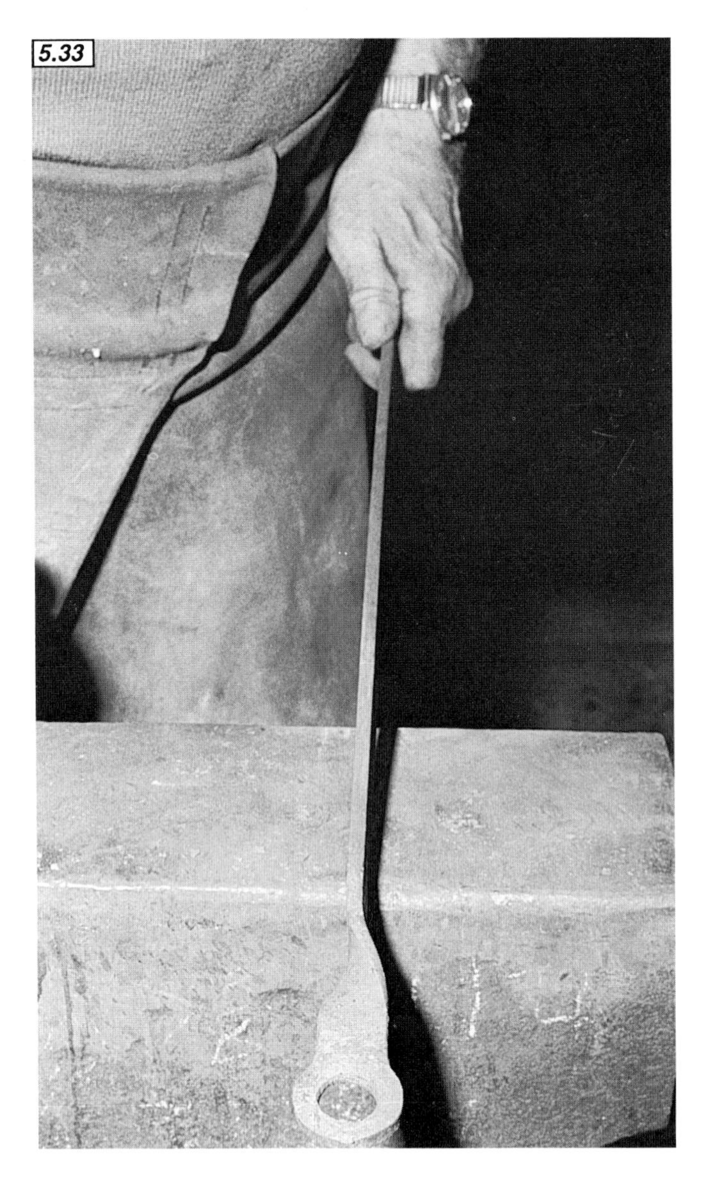

5.33 . . . leaving a neat and tidy ring.

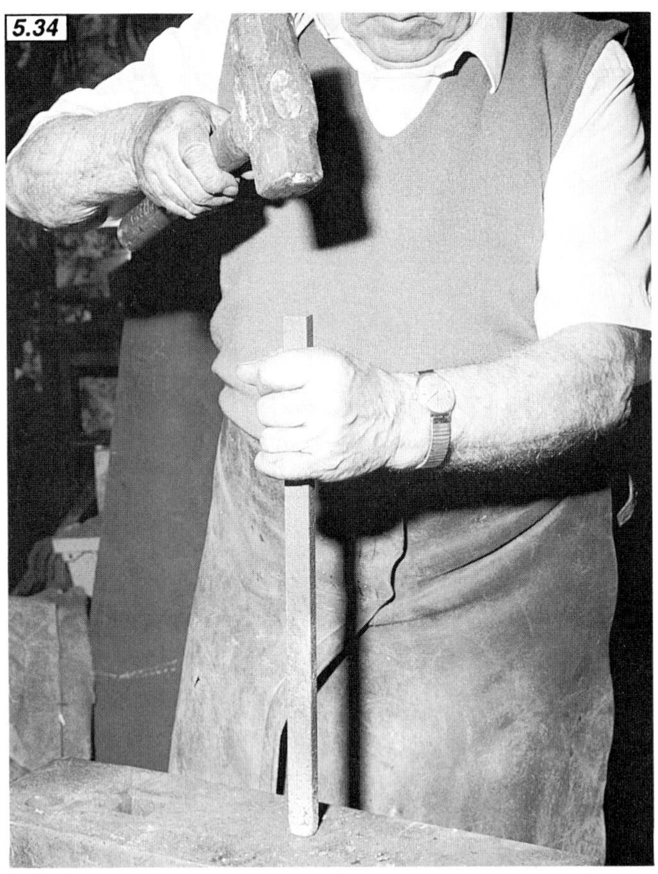

5.34 Where extra metal is needed for a particular operation, upsetting is the answer. Frank had brought the bottom few centimetres of this bar to a good working heat, and is using vertical blows from a big hammer to push metal up in a bulge. Short work can be accommodated on a normal-height anvil; Frank's forge has another set flush with the floor for lengthier jobs. A solid, non-resilient surface is needed for swift upsetting—a concrete floor isn't good enough. The process is repeated, taking fresh heats and keeping the work true as required.

By swelling (rather than removing) material, punching allows much bigger holes to be made than are possible with a drill. But as with all forge work, advance thinking is needed. If it looks as though punching will leave insufficient strength around the hole, thickness can be locally increased by *upsetting or jumping-up*.

This involves swelling metal locally. Figs 5.34 and 5.35 should make the process clear. Upsetting is a powerful weapon in the smith's armoury, and one which finds use in many facets of his work. It's often the first task in a sequence, carried out to create a reservoir of metal for future operations. And where the job allows—and that's often, as long as forethought goes into the order of attack—upsetting is a much better bet than laying down weld and grinding it back.

5.35 Plenty of thickness here for subsequent forging operations. By selective cooling as in Fig 5.31, the upset area could easily have been produced partway along the bar rather than at its end. It's an ideal way of providing extra material locally prior to punching a large hole.

Tired of all that welding?

Every repair job involves decisions. Not the least of these is allocating blame, or—more importantly—absolving yourself from it. This is often the fitter's most basic and pressing dilemma.

The handy cut-out flowchart below covers most eventualities; follow it through for a solution for any given situation. Tested in various industries (including farming) it hasn't yet failed.

(With thanks to the Institute of Diagnostic Engineers.)

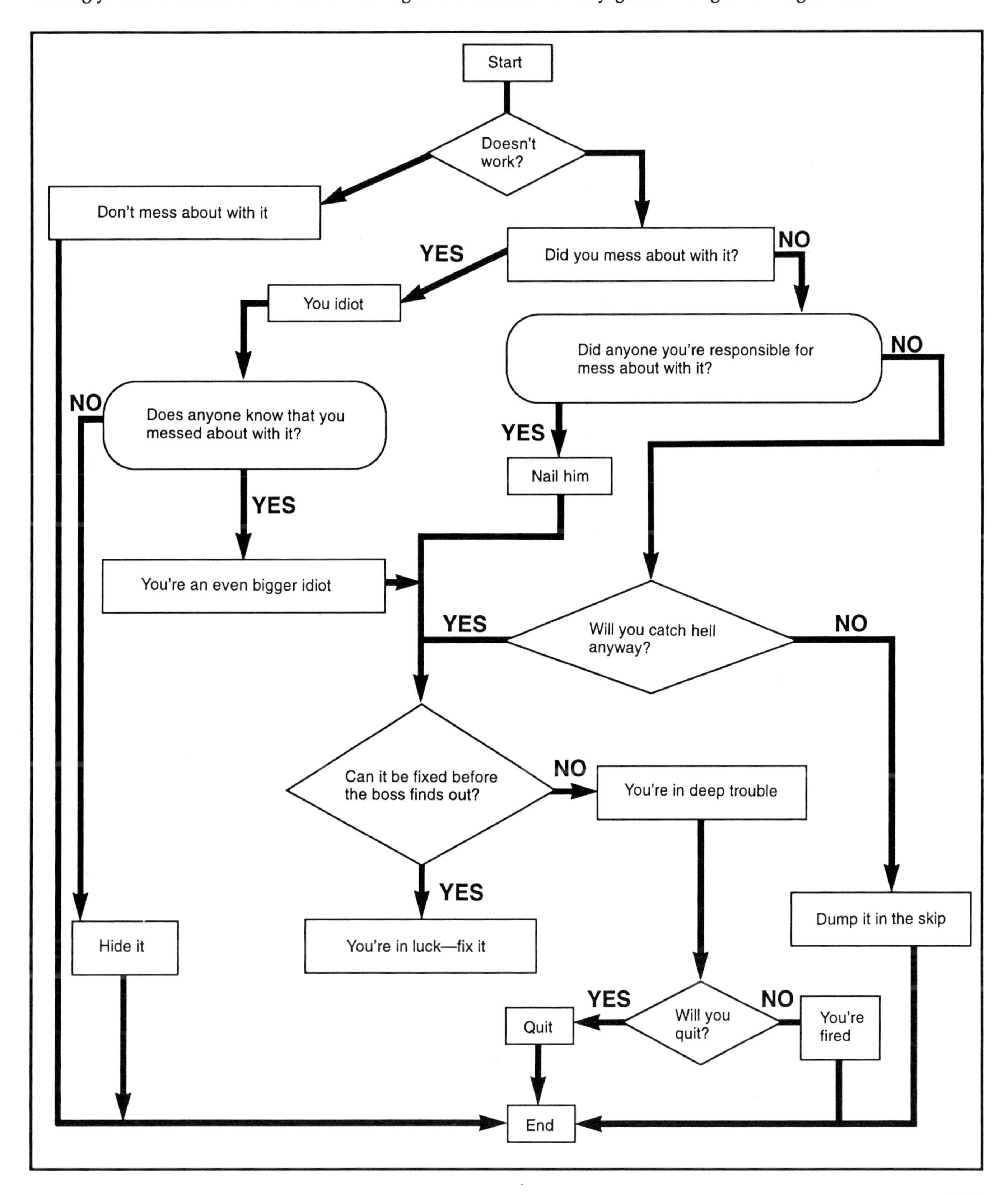

INDEX

INDEX

FARMING PRESS BOOKS

Below is a sample of the wide range of agricultural and veterinary books published by Farming Press. For more information or for a free illustrated book list please contact:

Farming Press Books, Wharfedale Road
Ipswich IP1 4LG, United Kingdom
Telephone (0473) 241122 Fax (0473) 240501

Farm Workshop BRIAN BELL
Describes the requirements of the farm workshop and illustrates the uses of the necessary tools and equipment.

Farm Machinery BRIAN BELL
Gives a sound introduction to a wide range of tractors and farm equipment. Now revised, enlarged and incorporating over 150 photographs.

Machinery for Horticulture
BRIAN BELL & STEWART COUSINS
A description of the basic functions and uses of the diverse machinery used in all aspects of horticulture.

Farm Building Construction
MAURICE BARNES & CLIVE MANDER
Details in blockwork, brickwork, timber, concrete, flooring, roads etc. for new buildings or improvements.

Tractors Since 1889
MICHAEL WILLIAMS
An overview of the main developments in farm tractors from their stationary steam engine origins to the potential for satellite navigation. Illustrated with colour and black-and-white photographs.

Fordson: the story of a tractor (video)
This colour VHS video features the five main Fordson models from 1917 to the 1950s. It combines archive material with new film.

Ford and Fordson Tractors and **Massey-Ferguson Tractors**
MICHAEL WILLIAMS
Heavily illustrated guides to the models which made two leading companies great.

Farming Press Books is part of the Morgan-Grampian Farming Press Group which publishes a range of farming magazines: *Arable Farming, Dairy Farmer, Farming News, Pig Farming, What's New in Farming*. For a specimen copy of any of these please contact the address above.